GL⊕BAL WARMING

HUMANITY MUST SURVIVE
THE 21ST CENTURY

Volume 1

Prof. Theodore Vornicu, Ph.D.

authorHOUSE®

AuthorHouse™ UK
1663 Liberty Drive
Bloomington, IN 47403 USA
www.authorhouse.co.uk
Phone: UK TFN: 0800 0148641 (Toll Free inside the UK)
*   UK Local: 02036 956322 (+44 20 3695 6322 from outside the UK)*

Published by AuthorHouse 08/16/2021

ISBN: 978-1-6655-9241-3 (sc)
ISBN: 978-1-6655-9242-0 (hc)
ISBN: 978-1-6655-9243-7 (e)

This book is dedicated
to my brother's family Vornicu: Petru, Mioara,
and their son Stefan

My brother picked me up from a hole and has given
me the light of Education.

Contents

PREFACE

The Coronavirus might be the biggest story of the decade, but the Climate Crisis will be the grand narrative of the century, if not for centuries to come.

I wrote this book to be a GUIDE for humanity about global warming.

I wrote it with the mind of a scientist, from the heart of an Earth citizen, and the soul of a worried member of humanity. The planet is facing its most challenging time in the history of humanity.

I am optimistic that my generation's decision-makers, in this decade 2021-2030, will prevent the point of no return for the climate crisis. This way their kids and grandsons and granddaughters won't blame this generation and will have a normal life on the *"Pale Blue Dot"*, maybe unique in the Milky Way galaxy, if not on the entire Universe. The generation of 2021, has to assure the basis for preventing the collapse of the warming planet, or other risks due to human activities. The young generation of today is expecting to reach the end of the century. They have to be part of developing science, innovation, and technology, avoiding the risks of catastrophic events, for themselves and the next generations. How they approach and make decisions in their lifetime will be crucial for the future of human life on this planet Earth. The current trend of destruction and extinction of life on the planet leads me to the theme of this book. I can not stay with my arms crossed, looking at the development of dramatic events without taking them to the public. I wish this book to be a guide for the public, to a greater understanding of global warming and its catastrophic impacts on humanity.

I dearly wish to emphasize that a handful of major newspapers are paying attention to global warming. They are the heroes of Climate Education and Action. Because, most news coverage, especially on television, continues to underplay the climate story, regarding it as too complicated, or disheartening, or controversial. Last April 2021, we: Covering Climate Now consortium - a group of news journalists CCNow, asked the *"world's press to commit to treating climate change as the emergency"* that scientists say it is; their response was dispiriting. We created CCNow in April 2019 to help break the media's climate silence. Since then, CCNow has grown into a consortium of hundreds of news outlets reaching a combined audience of roughly 2 billion people. The climate coverage of the media as a whole has noticeably improved. CCNow is an example for all affected by global warming in their action. Maybe the leaders of most carbon emitter countries need to take a flight to the International Space Station, to see from space our planet is in peril, and come back to act, so the *"planet Earth is not becoming a Mars planet in the Solar system."* I have to be grateful to those who encouraged me to write the book, my twin sister Maricica, the colleagues from Technical University of Iasi city, Faculty of Construction, Romania, especially Professors: Mihai Budescu, and Vasile Musat. I also need to warmly appreciate the moral support from friends: Marian Enache - Regensburg, Germany, Catalin Panaite - London. Profound recognition to my special friends and consultant: Thilee Subramaniam, Technologist, California.

"MORE PRAISE FOR "GLOBAL WARMING - HUMANITY MUST SURVIVE THE 21ST CENTURY"

In the second edition of his highly relevant book "Global Warming - Humanity Must Survive the 21st Century", Prof. Theodore Vornicu gives a broad perspective of the global warming problem. His coverage is complete, and he builds up the historic, economic, and political events that have brought up the current situation logically and engagingly. He also makes a compelling case for urgent action.

Prof Vornicu doesn't just stop at highlighting the urgency. He then goes on to provide information on the current and emerging "green energy" solutions out there. He presents a detailed analysis of the risk/benefits of these green technologies, their world economic impact, the economic opportunities, and the benefits they provide towards the health of the "Pale Blue Dot" that we live in. I'd like to see this book present in every library, used in the high school curriculum, and read by business leaders, politicians, and professionals all around the world.

Thilee Subramaniam, Technologist, California, USA

"Global Warming - Humanity Must Survive the 21st Century" is a delightful must-read for every responsible citizen on this planet. With originality, Prof. Theodore Vornicu addresses the most important of today's acute subjects of human extinction risks by human-made causes for catastrophic disasters, due to global warming. A must-read book for any conscious humans on our planet.

Prof. Mihai Budescu, Ph.D., BSc(Eng), Dept. of Structural Mechanics, Faculty of Civil Engineering "Gheorghe Asachi" Technical University of Iasi, Romania.

The book *"Global Warming - Humanity Must Survive the 21st Century"* is captivating, very actual, and is addressed to all citizens of the planet, who are concerned about the future of life on Earth. A guide for all Colleges and University students, all professionals around the planet.

Emeritus Prof. Vasile Musat, Ph.D., Dept. of Ways of Communication and Foundation, Faculty of Civil Engineering, "Gheorghe Asachi" Technical University Iasi, Romania.

The book *"Global Warming - Humanity Must Survive the 21st Century"* constitutes the basic principles for a course to be taught in Colleges and Universities Worldwide about Global Warming. A must-read book.

+Prof. Constantin Filipescu, Ph.D., Depart. of Mathematics, University *"AL. Ioan Cuza"*, Iasi, Romania.

"In his book, Mr. Prof. Vornicu does an exhausting study of Global Warming, including all aspects of this phenomenon. Besides discussing the scientific aspects of Climate Change, Mr. Vornicu is presenting the most important climate events of the last 18 months. Also treating the effect of pollution, financial risks of investments, the impact of climate change on society, as well as action required by humanity, to slow down the planet warming. this book is a wake-up for all citizens of the planet Earth, and constitutes on real guide for people to get used and understanding this silent killer phenomenon."

Prof. Alexandru Secu Ph.D., Technical University of Iasi city, Romania.

"Prof. Vornicu has written an excellent sequel to his first book about the future of humanity. This is a wake-up call (maybe the last) for our generation in relation to the actual climate problems and the destruction that humankind is causing. Through the deep insight into the COVID-19 Pandemic, the reader is also taken on a way of reflection about the ephemerality of our very existence. This must-read book is a plea to change in thinking and consideration for the next generation."

"Marian Enache, Public Servant of the Free State of Bavaria, Regensburg, Germany"

Prof. Theodore Vornicu has written a unique book about "Global Warming." It is a book for all living generations and generations to come. It explains the physical phenomena of climate change, giving the most recent weather dramatic events. Also, Prof. Vornicu analyzes fossil fuel developments. Detailed global primary energy (based on fossil fuels) and renewable energies constitute an ample review and presentation. Financial analysis of investments and bank loans to companies based on fossil fuels, to comply with the acute impact of climate change, are treated as mandatory action - dollar not only speaks but dictates. The magnitude of global warming upon humanity with catastrophic impact, and actions required by society to attenuate climate impact in the next decades, are exhaustively analyzed in the 6th chapter of the book. The conclusions are speaking and addressing the entire humans of the planet, as an emergency to change human behaviors, and rapid climate actions, to save the life on the planet for generations which are not yet born.

Ion Sococol, Doctorand Student, Technical University, Iasi City, Romania.

INTR⊕DUCTI⊕N

Instead of MOTO

People in San Francisco, and elsewhere in California, in late August, and the beginning of September 2020, woke to a deep orange sky that triggered apocalyptic visions in a year already rife with disturbing events. Skies so dark at times that it appeared more night than a day were accompanied in some places with ash falling like snow, the cause being massive wildfires filling the air with smoke, and cinders. The orange skies this morning are a result of wildfire smoke in the air, *"San Francisco Bay air quality officials said in a tweet. These smoke particles scatter blue light and only allow yellow-orange-red light to reach the surface, causing skies to look orange"*. As smoke gets thick in some areas, it blocks sunlight causing dark skies, the officials explained. Photos of the eerie scene, particularly of a San Francisco skyline fit for a dystopian science fiction film, spread quickly on social media.*" Is there a word for 'the apocalypse is upon us burnt sienna'?"* Others likened the scenes to planets other than Earth. *"If literal fire skies don't wake us up to a Climate Crisis, then nothing will. (Scientific American magazine announced on April 19, 2021, that it would stop using the term climate change in articles about man-made global warming and substitute* ***climate emergency")***. Enjoy joking about how crazy this year is because we made this mess and *"it's only going to get worse"*. Dark skies blocking the sun chilled temperatures at what has historically been the warmest time of the year in San Francisco." *Geo-color imagery shows a very thick multilevel smoke deck over much of California,"* the US National Weather Service said in a tweet. *"This smoke is filtering the incoming energy from the sun, causing much cooler temperatures and dark dreary red-shifted skies across many areas"*. What were being described as *"unprecedented"* wildfires, fueled by strong winds and searing temperatures, were raging across a wide swathe of California, Oregon, and Washington on Wednesday, Sept. 9th, 0220, destroying scores of homes and businesses in the western US states and forcing hundreds of thousands of residents to evacuate. In California, where at least *"eight deaths"* have been reported, National Guard helicopters rescued hundreds of people trapped by the Creek Fire in the Sierra National Forest. In Oregon. Governor Kate Brown declared the fires in the northwestern state to be a *"once-in-a-generation event"*. Jay Inslee, the governor of neighboring Washington state, described the wildfires as *"unprecedented and heartbreaking",* who campaigned for the Democratic nomination for president on a platform of tackling climate change, and California Governor Gavin Newsom both blamed the effects of a changing climate for the exceptional ferocity of this year's blazes. *"I quite literally have no patience for climate change deniers,"* Newsom said. *"It's completely inconsistent, that point of view, with the reality on the ground."*

The planet Earth is warming up. Humanity is experiencing more intense and more frequent weather phenomena all year around. One of the key questions the scientists face is: *"Could there be a connection between record cold, intense storms, and tornadoes amid global warming?"* The U.S. has endured a wild stretch of harsh winter weather in the middle of February 2021, in Texas, the Vortex. *"Thanks to an invasion of the infamous polar vortex".*

It may be counterintuitive, but **could global warming have caused this?** I wish to clear the answer to this question before advancing in more scientific details of global warming.

First, an explainer: The polar vortex is a gigantic cold circular upper-air weather pattern in the Arctic that envelops the North Pole. It's a normal, natural pattern that is stronger in the winter and tends to keep the coldest weather near the North Pole. The jet stream usually opens the polar vortex in and keeps it there. Some of them can break off, and move south, bringing extremely cold weather down into the U.S., Europe, and Asia.

"Not all the scientists say there could be a connection between global warming and the wandering polar vortex". The theory is that when weird warmth invades the Arctic, some of the colds slope down south into North America or Europe. *"There is evidence that climate change can weaken the*

polar vortex, which allows more chances for frigid Arctic air to slip into the Lower 48 parallel." University of Georgia meteorology professor Marshall Shepherd said. Woodwell Climate Research Center and climate scientist Jennifer Francis, who has published a study on the phenomenon, said in 2019 that *"warm temperatures in the Arctic cause the jet stream to take these wild swings, and when it swings farther south, that causes cold air to reach farther south."*

"Are February tornadoes unusual?"

Regarding the vortex in the South this week,(middle of February 2021) scientists say *"there is no clear connection between that type of severe weather and human-caused climate change."* While climate change does have a documented effect on many extreme weather events, it has no clear connection to severe thunderstorms nor the tornadoes they produce.

In fact, a 2016 report from the National Academy of Sciences, Engineering, and Medicine found that of all-weather phenomena, severe storms (and tornadoes) are the most difficult to attribute directly to climate change. On average, *"there are 29 tornadoes in February of each year across the nation,"* the National Oceanic and Atmospheric Administration (NOAA) said. Deadly tornadoes: 3 dead, at least 50 homes damaged as tornado rips through North Carolina (February 17-24, 2021). And overall, global warming remains real, despite this week's chill: *"The six years from 2015 to 2020 were Earth's warmest six years in recorded history"*, according to NOAA. *"Global temperatures have also been above average for 433 consecutive months." "Temperatures in the central U.S. near record lows this week"*, February 17-24, 2021. (More about this subject, see paragraph 2.2.5).

The planet is already experiencing a *Climate Crisis*, which will last for centuries to come, and in the year 2020, the *"Pandemic Crisis"* overlapped the climate crisis, creating a very dire situation for humanity. Also, this year 2020 has seen record-breaking heat, wildfires, and storms, continuing in earlier 2021 with an extreme *"Polar Vortex"* over South Central United States, especially in Texas. All these extreme weather events are presented in detail in chapter 2. *U.N. Secretary-General António Guterres urged every country to declare a "climate emergency." He also declared Global Climate Emergency, until green gases emission rich absolute net-zero"*.

Globally Public Backs Climate Change Actions.

The biggest global survey of its kind has found that almost 64% of people believe climate change is a global emergency. The *"Peoples' Climate Vote"*, published January 27, 2021, conducted by the United Nations Development Programme (UNDP) and the University of Oxford, surveyed 1.2 million people across 50 countries. In total, 64% of respondents agreed that climate change is a crisis for centuries to come. The survey also found that *"the majority of young people are more concerned about climate change. The survey found 70%t of young people called climate change a global emergency compared to 58% of adults aged over 60"*. Since 2018, millions of students from Southeast Asia to Washington are demonstrating to demand global action on the climate crisis. The Fridays for Future movement, which began as a lone demonstration by Swedish teen Greta Thunberg, quickly expanded, with millions of students taking to the streets to *"push the climate crisis to the top of the agenda of world leaders."* In Britain, making companies pay for carbon and methane pollution had huge support while the majority of those asked in the United States backed green energy sources. Also, in the United States, young people agree to pay a *"Planet Tax of $20. per month, to know that their kids will enjoy a secure, safe and clean habitat"*. Investing in green jobs and more climate-friendly farming techniques were popular in Indonesia and Egypt, the survey found. *"The Peoples' Climate Vote has delivered significant data on public opinion that we've never seen before,"* said Professor Stephen Fisher from the University of Oxford. *"Recognition of the climate emergency is much more widespread than previously thought."* The world is not only facing a Covid-19 emergency but also *"security, and existential threats"* to human habitat, and biodiversity, and political stability between big power states like Russia, China, and the USA. News January 27, 2021, By: Adela Suliman

World leaders' 10 biggest fears for the decade 2021-2030

The coronavirus crisis should spur the world to *"wake up to long-term risks"* such as climate change, collapsing asset prices, and weapons of mass destruction, according to the World Economic Forum (WEF). Here we must include, as the climate crisis impacts: migration, population increase, security, scarcity of food and water, and the global arms race.

Global leaders' biggest concerns about the decade ahead are laid out in a new report by WEF, best known for organizing an annual summit of leaders at the Davos ski resort in Switzerland. More than 800 business, government, and civil society leaders were surveyed in 2020 about perceived risks in years to come for the WEF Global Risks Report 2021, published on January 19, 2021. Respondents were asked to rank 35 potential short-, medium- and long-term risks. *"Infectious diseases and livelihood crises"* were the two greatest concerns for leaders over the next two years, in a sign of the severe and lasting damage expected to result from COVID-19 and government curbs on economies. Other *"clear and present dangers included extreme weather and digital inequality"*. WEF's report warns *"COVID-19 risked widening the gap between the haves and have-nots"* in both access to technology and digital skills. Over the second half of this decade, the biggest concerns include: *"weapons of mass destruction, state collapse, biodiversity loss, adverse tech advances, and natural resource crises"*. Leaders were also asked to *"highlight the risks they believed would be most damaging"*, as well as those that were most likely. *"Climate change and the environment loomed largely"*, but leaders appeared *"pessimistic about humanity acting successfully to address their biggest concerns."*

"Failure to take action on climate change will have the largest impact of any issue." Extreme weather, failure to act on climate change, and human environmental damage were also perceived as the three most likely risks", according to the survey. As governments, businesses, and societies begin to put the pandemic behind them, they have to shape new economic and social systems. That will improve our collective resilience and capacity to respond to shocks while reducing inequality, improving health, and protecting the planet. Due to the mess Capitalism created on the planet, the survey included 12 new risks compared to the previous report. They include the collapse of systemically important industries, social security systems or multilateral institutions, deteriorating mental health, mass youth disillusion, prolonged economic stagnation, fracturing international relations, and a *"pervasive backlash"* against science. (Zhurnal Zekirija Haxhiu Tuesday, January 19, 2021). Technology also appears a growing concern, with digital inequality, digital concentrations of power, and

failure of technology governance all added in 2020 (1). Covid clearly showed that the capitalist style of *"people living from pay-day to pay-day"* is not anymore the "Fundament of Capitalism". For instance, in an existing global climate crisis at the beginning of 2020 humanity started to face a health pandemic crisis, which was followed by economic, and mental health issues. The US, on top of these crises, has developed a social and political crisis. On January 6, 2021, the political crisis reached an apogee, fortunately ending this crisis, showing that America is experiencing a public education crisis too, as scientists in the field, even army generals, recognize it. This crisis is not only on the public level but even to the top of the political elite, showing that the principle: we do education based on IQ, and gain conscience based on education, is not correctly applied (6).

A Decisive Election For Humanity.

During the climate crisis that will last centuries (like the little ice age 1400-1800 AD), a superpower like the United States with enormous influence on global relations among all nations, the presidential elections determine the character of the country for the next four years. And they have a lot to say about what the world will feel like too. *"But the election of 2020 may determine the flavor of the next four millennia. That's because "time is the one thing we can't recover, and time is the one thing we've just about run out of in the climate fight."* The Intergovernmental Panel on Climate Change (IPCC) in its 2018 report made it clear that we, **HUMANITY**, have until 2030 to make *"fundamental transformations in our energy system"* reducing carbon emission to 50%. Read that sentence again. Because it *"carries deep political implications"*. Very few of the problems that the government deals with are time-limited. Issues like housing or education or healthcare last throughout our lifetimes. The Climate Crisis is totally different. If we don't slow it down soon, we will not solve it because we will move past tipping points from which we have no retreat (See paragraph 2.1 for tipping points). Some we've passed already: the news that *"Greenland is now in an irreversible process of melt, Antarctic ice melt, and heating of the Atlantic ocean water surface"*. The Trump presidency cost us dearly. The Paris climate accord was completely undercut by the administration's insistence on rolling back environmental laws, favoring the oil and gas industry, and removing the US from international negotiations (4 dark American years). In the last few weeks, we've watched the Chinese make new pledges at the UN session in Sept. 2020, and the state of California announced a prospective end to the era of the internal combustion.

A *"Biden administration"* can join in those efforts; indeed it can lead them. Vice president Kamala Harris has announced that one of her first acts would be to convene a meeting of high-emitting nations, in anticipation of the next UN meetings in Scotland in November 2021. Four more years of Trump and all-out climate denial there will be no way to put any kind of pressure on leaders like Russia's Putin or Brazil's Bolsonaro. The effective chance envisioned in the Paris Accords will die forever. And the *"job of future presidents will increasingly involve responding to disasters that it's no longer possible to prevent."* The 1.3C that we've already increased the planet's temperature has taken us into what is effectively a new geological era, one markedly less livable for humanity. But it still bears some resemblance to the world that our civilizations emerged from. If we value those civilizations then *"a vote for Joe Biden isn't really about the next four years. It's about to save the humans on this planet* (2). Thank God and Providence that happened.

Biden Presidency, and Climate Change.

President Joe Biden's attitude and policy agenda are totally opposed to the Trump administration on everything. The most drastic difference between them is the climate crisis. Trump, who has regularly questioned the science of the climate crisis, because he is parallel with science, has rolled back more than 100 environmental regulations and is working to eliminate dozens more before leaving office. Biden, on the other hand, has called the climate crisis the *"Number 1 issue facing humanity, his environmental policy agenda is more ambitious than any president in American history."* Biden's expansive climate plan includes massive investment in green technologies to move the U.S. to *"100 percent clean energy by 2035 and reaching net-zero emissions by 2050".* His *"Build Back Better"* economic agenda relies on creating new jobs in green infrastructure. Biden has named a climate team of veteran lawmakers who share his progressive environmental vision.

John Kerry as Biden's climate czar, a former senator, presidential nominee, and Secretary of State, John Kerry is set to begin as *"first-ever special envoy on climate."* Kerry's new position will address what the president called an *"existential threat"* for humanity. *"For the first time, the United States will have a full-time climate leader to participate in ministerial-level meetings, and that's a fancy way of saying he'll have a seat at every table around the world. For the first time, there will be a principal on the National Security Council who can make sure climate change is on the agenda in the situation room,"* Biden said in Wilmington. *"He will be matched with a high-level White House climate policy coordinator and policymaking structure that will lead efforts here in the United States to combat the climate crisis, mobilize action to meet the existential threat that we face,"* Biden said. *"Let me be clear: I don't for a minute underestimate the difficulties of meeting my bold commitment to fighting climate change, but, at the same time, no one should underestimate for a minute my determination to do just that."* Kerry's most recent federal government position was secretary of state under President Obama. In that capacity, he helped write the Paris climate agreement, by securing commitments from nations to reduce carbon emissions. The majority of climate scientists have urged political leaders to take quick action to solve the climate crisis. After being introduced in Wilmington, Kerry said of climate change that *"no country alone can solve this challenge. Even the United States, for all of our industrial strength, is responsible for only 13% of global emissions. To end this crisis, the whole world must come together. You're right to rejoin Paris [climate agreement] on day one, and you're right to recognize that Paris alone is not enough,"* Kerry said. *"At the global meeting in Glasgow, all nations must raise ambition together or we will all fail together, and failure is not an option. Succeeding together means tapping into the best of American ingenuity, creativity, and diplomacy, from brainpower to alternative energy power, using every tool we have to get where we have to go."* Like Biden, Kerry said we must fight against climate change as an *"exciting"* opportunity," saying *"it means creating millions of middle-class jobs, it means less pollution in our air and our ocean, it means making life healthier for citizens across the world, and it means we will strengthen the security of every nation in the world."* Biden said about Kerry: there was *"no one I trust more. Fifty-seven years ago Joe Biden and I were college kids when we lost the president who inspired both of us to try to make a difference, a president who* reminded us that here on earth, God's work must truly be our own," Kerry said, in a reference to former *"President John F. Kennedy"*. President Joe Biden *"will trust in God, and he will also trust in science to guide our work on earth to protect God's creation."* (3). More about President Biden's climate program is presented in paragraph 6.2.

Climate Race for Global Supremacy

Put aside the trade and technology wars. *"The next race for global political and economic supremacy will focus on climate."* That's according to Bank of America Corp.'s research group, which said *"climate change will be this decade's most important theme",* just as technology underpinned economic growth during the past decade. Haim Israel from Bank of America said *"China has spent twice as much as the U.S. on climate action. We believe climate strategies offer a route to global supremacy,"* he said. *"Whether through regulation, limits on exports, tariffs or significant investments, we believe the U.S. and China will do whatever it takes to take the lead on climate action."* Based on the current situation with carbon emission, China will continue emission for more than at least 4 decades, which will be catastrophic. China is emitting almost one-third of the global emissions of greenhouse gases (GHG).

"Bank of America estimates that climate change could reach $69 trillion this century," and *"investments in the energy transition need to increase to $4 trillion a year".* Only in research innovation will be more than *"$100 billion a year.* Bank of America estimates that the *"potential market capitalization for companies tackling climate change to be about $6 trillion annually.* Most important in the climate war include supply-chain dominance, domestic-focused manufacturing policies, human rights-related laws, and carbon-related footprints, and trade tariffs. Less foreign energy dependence and a focus on technology exports could be key. China will increase wind and solar capacity to three and four times, respectively, by 2030. That compares with the U.S., where wind capacity is set to double and solar is set to grow three times. The EU will produce more green energy than China (4. Bloomberg, Feb. 8, 2021, By Seijel Kishan).

Climate scientist Michael Mann

"The U.S. is in position to be a leader on climate change."
Well-known climate scientist and Penn State professor Michael Mann argues in a new book that there are many tools for addressing climate change and transitioning to green energy.

Mann's book, *"The New Climate War: The Fight to Take Back Our Planet,"* comes as the country's climate policy is expected to change with the Biden administration. Mann said he thinks the *"U.S. is in a position to re-establish its global leadership on this issue"*, and that Joe Biden has indicated he will do that as president. *"He campaigned on the climate issue, on the climate crisis,"* Mann said. *"He has the mandate to act on the climate crisis, and he is incorporating climate action into every single sector of our executive branch now."* An important action is needed by 2030 to slow down as much as possible the effect of the climate emergency. Mann argues in part the U.S. should use a carrot and stick approach, subsidizing renewable energy and using carbon pricing, *"such as a carbon tax or cap-and-trade policy"*, to move the country away from fossil fuels. *"You can do that by leveling the playing field so that renewable energy, which isn't degrading the planet the way fossil fuel energy is, can compete fairly in the market. We have to use the tools that are available to us within the context of the existing economy,"* Mann said. *"Carbon pricing is a big part of that."* Mann, a distinguished professor of atmospheric science who has served on scientific committees to report on climate change, said climate change can't be denied anymore. He says the *"fossil fuel industry is moving away from discrediting climate science. Instead, it's using tactics like focusing on individual behavior rather than policy changes, to protect its interests through denial methods. The impacts have become so obvious that they can't deny that climate change is real or human-caused anymore, so they've turned to more insidious and nefarious means of trying to distract the public and block action,"* he said. I wish to add that: *"even if all emissions were zero tomorrow, heating, droughts, floods, storms, and rising sea levels will just continue"* (5). This short presentation in the introduction, emphasizes the importance of climate change for humanity, under all aspects of life on the planet. Today, generations and generations to come will live with climate change, and based on our actions, how we can manage the slowdown of this crisis, the future generation will live. *"Our actions are detrimental for the life of those who are not yet born"* (6). *"I think the next generations are screaming at us to fulfill our promise"* (John Kerry).

The Content of the Book.

Following will be a short presentation of the content of the book (6 chapters).

Chapter 1 I make a succinct presentation of science and technology development 19 and 20 centuries, development which triggered climate change, by using fossil fuel to supply the hungry industries and society with power. From water vapor turbines to Apollo and artificial intelligence (AI), was a long way to go in a short period, if compared with the previous 20 centuries.

Chapter 2 is dedicated to the science behind global warming, and the most turbulent weather atmospheric events of the years 2019 and 2020, which have been the warmest years in modern time. One of the most important for climate and ecosystem is the tipping points. If they are reached then the weather events grow in intensity and frequency to a catastrophic level.

A 2019 UNEP study indicates that now *"at least for the Arctic and the Greenland ice sheet a tipping point has already been reached. If damaging tipping cascades can occur and a global tipping point cannot be ruled out, then this is an existential threat to civilization. No amount of economic cost-benefit analysis is going to help us. We need to change our approach to the climate problem."* Paragraph 2.1.2 sounds like conclusions for climate science. The scientific data are detrimental in drawing the correct conclusions. The weather phenomena are the strongest defenders of global warming, so I will make a summary of events. For years, climate scientists have been wary of attributing extreme weather directly to man-made atmospheric warming, but that's changing in the face of historic heat waves and cascading natural disasters. In the summer of 2020, *"derecho"* a complex of unusually powerful, hurricane-like storms, tore through the Midwest, destroying homes and crops across a 745-mile path (1,192.Km); Hurricane Laura crashed into the Gulf Coast with sustained 150-mph (240.Km/h) winds. In California 7600. of wildfires incinerated an area of 2.5 million acres, (860,000. hectares) the size of Rhode Island in just a week. The Southwest suffered a punishing heatwave with a high of 130F (F stand for degree Fahrenheit), 54.4 C, in Death Valley CA., perhaps the hottest day in world history. *"It followed highs of 125F (49C) in Iraq and a record 100F (38C) in the Siberian town of Verkhoyansk, a once-in-100,000-years event."* These are the result of *"human activities"* pumping 2.6 million pounds of CO_2 into the atmosphere per second, (1,182. tons of CO_2 / second = 37.27 billion tons/year). *"We've gotten to the point where, when it comes to extreme heatwaves, there is almost always a human fingerprint,"* said UCLA climate scientist Daniel Swain. The expression *"500-year storm"* is losing its meaning: Houston has suffered five of them in five years. California's *"wildfires ignited by 11,200 dry lightning strikes in 72 hours"* produced the second-and third-worst blazes in state history, even without the aid of the fall's strong Santa Ana winds. The Atlantic coast has seen 30 named storms, 12 of them qualifying for a hurricane. *"Hurricanes have done $335 billion in damage over the past three years, compared with $38.2 billion across the entire 1980s, adjusted for inflation"*. Climate disasters of all types inflicted $807 billion in damage during the 2010s, the hottest decade on record. Weather patterns are shaped by an intricate web of atmospheric and oceanic conditions. But when both rising temperatures and disasters become consistent and pervasive, the connection becomes obvious. *"The average daily highs in Northern California during wildfire season are 3C to 4C degrees warmer than they were in 1900."* Warming of the planet's surface causes atmospheric instability that can produce stronger, more frequent storms while rising ocean temperatures (another tipping point close to being reached). The future of climate chaos is being previewed in northern latitudes, where a CO_2 domino effect plays out. Warm winters melt more snow, causing the ground to absorb more heat, which leads to dry soil that fuels wildfires and *"thaws permafrost, releasing carbon into the atmosphere."* In Russia this summer 2020, thawing permafrost caused a power-plant fuel tank to collapse, spilling more than 20,000 tons of diesel into the Ambarnaya River. *"Russia's average temperature was nearly 11 degrees F (6C) above its January-to-April norm"*, the largest anomaly ever for any country. In February, Antarctica hit a record 69F (19.5C), causing a 120-square-mile chunk of a glacier to break off. Disrupted weather patterns are rippling around the globe, creating bizarre, almost biblical catastrophes. Extreme temperatures in the Indian Ocean caused drought and wildfires in Australia while spawning cyclones in eastern Africa. The torrential rain there created perfect conditions for desert locusts, which reproduced at terrifying rates. By March, hundreds of billions of finger-length insects swept across the region, devouring every crop in their path, and pushing tens of millions of Africans to the brink of starvation. People are even experiencing climate change through their sinuses. Airborne pollen increases as temperatures climb, which is why residents of *"Alaska, where warming is happening twice as fast as the global average"*, report especially bad allergies. *"There's irrefutable data,"* said Jeffrey Demain, director of an Alaskan allergy center. Much depends on the oceans, which play a critical role in absorbing CO_2 and heat and regulating weather. *"The amount of heat we have put in the world's oceans in the past 25 years equals 3.6 billion Hiroshima atom bomb explosions,"* said Lijing Cheng, a Beijing physics professor. Warming oceans are circulating more slowly by about 15% in the Atlantic Ocean since 1950. The reduction in their moderating influence could cause warmer summers, colder winters, changing rainfall patterns, and more destructive storms. Climate change is no longer a theoretical threat. In California, average temperatures have climbed 1.8F degrees since 1980 while precipitation has dropped 30%, doubling the number of extreme-risk days for wildfires each year. Rancher Taylor Craig drove for his life as flames raced toward his Northern California home. Later, in the summer of 2020, sitting in a Walmart parking lot, Craig said he realized he had joined a new and growing club. *"I'm a climate refugee,"* he said. The pandemic forced automobile and airplane travel to fall off a cliff, and satellite images of pollution in the atmosphere offered a striking before-and-after contrast. At the height of April's coronavirus lockdowns, Google's mobility data indicated that *"4 billion people cut their travel in half."* As a result, worldwide daily CO_2 emissions dropped by an estimated 18.7 million tons, falling to levels not seen since 2006. Reduced car, bus, and truck traffic contributed to 43% of the drop-off, although emissions from residential buildings ticked up 2.8%, mostly from people running air conditioners while stuck at home. Scientists, however, are not celebrating. They anticipate just a 7% decline in carbon emissions this year 2020. By December the emission was back at the level of pre-pandemic. *"It goes to show just how big the challenge of decarbonization is,"* said Zeke Hausfather, a climate scientist at the University of California, Berkeley. To reach the global emissions targets of the 2015 Paris climate accord, CO2 would need to drop constantly, as it did in 2020 every year for the next decade. *If so the quantities reduced by 2030 will be just ⅓ from the 2020 level, which is not enough.* What about the existing carbon dioxide in the atmosphere (1.8 trillion metric tons) which has stood for more than 100 years? This 7% reduction is a pie in the eye to hide the real truth about decarbonization. The emission of carbon in 2020 will increase the total existing carbon in the atmosphere by 20 Gigatons of CO2 (if half is absorbed by ocean water, which is already at a very high degree of acidity). The issue is that total CO2 in the atmosphere is increasing from year to year, instead of decreasing. Is important to mention that the number of greenhouse gases existing in the atmosphere before 2019, triggered the climate events in the last 2 years (2019 and 2020) presented in paragraph 2.2. You don't need to be a rocket scientist to understand what would happen to the weather events if each year added more greenhouse gases? The reduction of 7% each year (from 20Gigatons) for the next 10 years would not be of any improvement if we take into consideration that the global industrial development rate will be between 5% and 8.5% in this decade.

Chapter 3. Fossil fuels as of 2019, and pollution. Fossil fuel contribution in global primary energy is as follows: Oil 33%, coal 27%, natural gas 24%, totaling 84%. The other sources of energy are hydropower 6%, renewable 5%, and nuclear power 4%.

Oil. The real global picture of the oil industry for the future looks catastrophic for global warming. All big explorers and extraction of oil have detailed plans to extract more oil. A few examples: Russia invests $111 billion in a new oil field in Artic, building 3 airports, 15 industry towns, creating 130,000. jobs. The field reserves are 5 billion tonnes of oil. Africa in the new oil basin Kavango to explore 12 billion barrels of oil and 3.26 trillion cubic meters of natural gas. Libya to increase the production of oil to 2 billion barrels per day. Guyana (South America) to produce by 2025, 750,000. barrels per day. Norway to extend 136 new oil explorations in the Barent Sea.

China to extend offshore rings of oil in the South China Sea, and import from Iraq 4 billion barrels per month for 5 years. India to double refinery capacity from 200 million tons to 450-500 million tons in the next 10 years. In 2021, demand for oil increases 10-15 million b/d above 2019 demand and may reach 115-120 million b/d. The oil peak is estimated to be 2030, but for OPEC countries to be 2040. The oil picture is gravelly.

Natural Gas. Natural gas consumption increased by 78 billion cubic meters (bcm), to reach 24.2% in global energy production. The most increase is in the USA, 85 bcm, followed by Australia and China. Saudi Arabia to invest $110 billion in new fields, so in 2024 to start the exploration of 2.2 bcf of gas daily, and 450mcf of ethane, 500000. bpd gas liquids. The EU is preparing for more gas consumption, wishing to get a deal with Russia for 2 pipelines of gas to Germany. At present time Asia is importing 340bcm of gas by 2040 will double it.

Coal. China has the largest consumption of coal in the World. China will add a new 148GW of power plants, almost equal to the EU, 149MW. The coal power stations are China 987. EU 149, U.K. 12. China is experiencing an annual growth of 8.5% for electricity. This means in 10 years, by 2030, China's electricity demand will more than double today's consumption.

In the USA the coal consumption is going down. Sweden and Belgium shut down the latest power plant before 2017, Austria, and Spain before 2021. Portugal will shut down their coal power plant by 2027, and Germany by 2038 (too late for climate crisis). Asia is addicted to coal, only China consumes 50% of global consumption.

Emission of CO2. The largest emitters of GHG are: China 29.34%, USA 13.77%, EU 9.57%, India 6.62%, Russia 4.76%, Japan 3.56%, Germany 2.15%, South Korea 1.52%, Iran 1.81%, Saudi Arabia 1.72%. First, 20 big fossil fuel companies are emitting 35% of total emissions.

The methane and carbon dioxide emissions and sequestration in agriculture are in detail presented. One aspect is to practice regenerative agriculture, to hold the carbon in the soil.

Generative agriculture in the USA may cost $15 to $20 per tonne of carbon, much lower than CCS (carbon capture systems), which goes from \$94 to \$232 per tonne. The agriculture sector needs a *"national plan to roll out a regenerative farming practice"*, and this includes bureaucratic negotiations, discussing new industry regulations, and creating accessible vocational training.

It would cost the USA about *"\$57 billion to convert another 1 billion acres by 2050, through regenerative practices, but doing that some 23.15Gigatons of CO2 could be sucked out of the atmosphere, globally. This gets to \$2.46 per ton of CO2."* These are fundamental steps to *"curtail emissions from buildings, which account for nearly 40% of the greenhouse gases the U.S. sends into the atmosphere"* (this percentage varies from country to country). What is needed to do with existing and new buildings to reach Paris Agreement requirements are presented in this paragraph. Housing, commercial and industrial buildings consume between 28% to 40% of global consumption. Another sector with high consumption of energy and release of pollution is transportation, which rates between 22% and 28%, depending on the country. Aviation consumes 3.5% and shipping 3% of global consumption. The issue is to reduce it to net-zero by 2050. Solutions are using electric vehicles (EV), or hydrogen-based consumption. Here are a lot of debates, from technology to innovation and politics. All aspects are presented in this paragraph. Heavy industries, like steel, cement, aluminum, petrochemical, and manufacturing are big consumers too. The rates are high, between 5% and 10% for steel, and an average of 5% for cement and aluminum.

Deforestation, Permafrost, and Peatlands.

Rainforest, Globally. This paragraph is of maximum importance in global warming, due to the amount of carbon dioxide and methane released in the atmosphere, or sequestered in soil and trees. *"The world's dependence on coal, farming, soy, palm oil, and mining has resulted in two-thirds of Earth's tropical rainforests being destroyed,"* and the remaining ecosystems being put *"closer to a tipping point."* Tropical rainforests once *"covered 14.5 million square kilometers, 13% of Earth's surface"*, according to RFN, but now, just *"one-third of that remains intact"*. Of the original area tropical rainforests once occupied, *"34% is completely gone, and 30%*

is suffering from degradation". All that remains is roughly 9.5 million square kilometers, *"and 45% of that is in a degraded state."* Only *"one-third of the original tropical rainforest is still intact."* Anders Krogh, who authored the report and works as a special advisor at RFN, said in a press release that *"the findings are alarming. Humans are chopping these once vast and impenetrable forests into smaller and smaller pieces, undermining their ability to store carbon, cool the planet, produce rain and provide habitats".* The **Amazon** forest, which provides 20% of planet oxygen, the largest in the world, is suffering from the highest rates of deforestation and fire. Only in May 2020, has been lost to deforestation. (230 sq. miles) 12% of the area, more than the same month one year earlier. This happened during the reign of president Bolsonaro, a climate change denial and criminal, and future accused of *"Crimes Against Humanity".*

The US President Biden proposed an *"International fund of $20 billion to save the Amazon"* forest. Tropical forests hold closely to 50% of terrestrial carbon stock. *"There is no solution to the climate, and biodiversity crisis without ending deforestation."* In the same situation are Southeast Asia (Indonesia and Malaysia) and Africa (Congo) rainforests.

*"**Peatland** cover approx. 3.7 million sq. km. areas larger than India, and store approx.*

415 Gigatonnes of GHG. 11 times the human activities released in 2019 (37.7 Gigatonnes)". The only way to stop permafrost thaw is to limit global warming. For instance, the temperature soared 10C (18F) above average across much of permafrost in Siberia in June 2020. Wide fire in 2020 engulfed much of the permafrost in Russia, *"releasing as much carbon in a month as total human CO2 emission in one year."* Now the reader understands the importance of permafrost in global warming, which for decades has been neglected by science. Degradation of peatland by human activities accounts for 5% to 10% of global GHG release.

Carbon Dioxide (CO2) Concentration.

According to the US agency (NOAA) the monthly average CO2 concentrations, recorded at the Mauna Loa Observatory in Hawaii, were 417.1 parts per million (ppm) this year compared to 413.33 ppm in April 2019. **It's the highest concentration since records began in 1958.**

CO2 concentrations a decade ago were 393.18 ppm, NOAA reported, an increase of 23.92 ppm. Not only are CO2 concentrations increasing but they are also accelerating, according to the data. *"During the 1960s, the increase over one year was an average of 0.9 ppm which has risen to an average 2.4 ppm a year in the past decade"*. NOAA says it is *"confident"* that the CO2 measurements *"reflect the truth about our global atmosphere and human activities"*

Sadly, fossil fuel companies, some of the biggest offenders as far as carbon emissions go, are conspicuous by their *"absence on this list of more than 100 companies"* that have committed to lowering their carbon footprint mostly using carbon capture and other technologies.

"The world needs to lower annual emissions by 29-32 Gigatonnes of equivalent carbon dioxide (CO2e) by 2030 to have a fighting chance to stay

below 1.5°C and slow down global warming. "That's ~5x (approx. 5 times) the current commitments by companies, organizations, and governments". We need to lower our GHG emissions *"by 48% over the next decade if we are to avert catastrophic planetary changes."* This is about the emission of CO2. All aspects of the carbon price, carbon footprints, carbon taxes (between countries), as well as political implications are presented in detail. The last paragraph of this chapter 3 is dedicated to **pollution** and the impact of pollution on human health and the cost of society.

The cost of 130 million in Europe annually rose to $190 billion (166 billion euros).

China saved 10000. lives during the lockdown in 2020, due to pandemic, when pollution went down too. Without pandemics in China, 48666 people are dying annually due to pollution. Besides actual issues with pollution China needs to take action regarding EV battery disposal. Worldwide, *"over 6.7 million people died in 2019, due to pollution."* Big cities like Lagos, New Delhi, Shanghai, Chicago... the west coast of the United States, encounters extreme pollutants. Oceans, seas, rivers all over are polluted, and present a threat to biodiversity. Artificial Intelligence started to help with software to better monitor the pollution and have proof to fight the big fossil fuels company cynicism.

Chapter 4. Any business is performed through the bank and with investments. The most developed banking and investment system started developing in the UK. Today the USA has one of the most adverse systems. At the same time the U.S. banking sector is far more exposed to the *"systemic and financial risks of climate change"* than previously understood by investors or disclosed by the banks, new research finds. This exposure is so significant that it could trigger a financial crisis, with more than half the syndicated lending of *"all major U.S. banks being exposed to significant climate risk."* The banks are mostly private, big fossil fuel companies are mostly private sector. How these two institutions are collaborating these days, may decide the future of climate change. Due to the old saying *"the money talks"*, the eventual death of oil and thermal coal won't come from environmentalists, or even directly from renewable energy, *"it will come when big banks decide to stop financing it, rendering it unbankable and unsustainable".* That's exactly what *"**Goldman Sachs** has just done, in a* *first for a major financial institution. Goldman Sachs"* is the first big U.S. bank *"to rule out financing new oil exploration or drilling in the Arctic, as well as new thermal coal mines anywhere in the world"*. The bank's latest environmental policy declares climate change as one of the *"most significant environmental challenges of the 21ˢᵗ century"*, and has pledged to help its clients manage climate impacts more effectively, including through the sale of weather-related catastrophe bonds. This is one point why I give more data in this chapter about banking and investments, which has to go through ESG requirements (see chapter 3 for more details on ESG). *"Former Bank of England Governor Mark Carney said one of his main focuses is achieving clarity in the market for carbon offsets, which can be built into something much larger to help achieve net-zero global emission targets."* One of these huge investment corporations is BlackRock. What is more important, BlackRock might soon decide to offer investors equity funds that target higher environmental standards, providing a big financial boost to those firms. BlackRock's investment actions will be initially incremental, baby steps. However, at more than *"$7 trillion in assets, BlackRock is so large that even small steps make a big impact."* The BlackRock CEO's letter is more evidence that investment policies do change, maybe sooner than later. And that is not good news for the fossil fuel industry. Climate Action 100+ aims to curb greenhouse gas emissions for investors. One of the world's leading investor groups pushing for more corporate action on climate change said *"it has added Saudi Aramco, the world's largest oil producer, to its list of target companies."* Climate Action 100+ (CA100+), the *"450 members of which manage more than $47 trillion in assets,"* seeks to engage with the companies responsible for much of the world's greenhouse gas emissions to encourage the transition to a low-carbon economy. Specifically, it wants *"each company to commit to emissions cuts in line with the Paris Agreement"* on climate change. The group has achieved some success with its strategy, including at oil majors Royal Dutch Shell and BP. Most important at this stage are required financial policies by governments, to work together with fossil fuel corporations and financial institutions against the impact on climate change.

President **Joe Biden** plans to use every tool at his disposal in the fight against climate change, *"including financial regulation"*. While not an intuitive choice, supporters say mandating that public companies and investment firms quantify and disclose climate risks, and the costs associated with them, is a bold step that could make ESG data as commonplace in corporate financial reports as sales and profit figures. *"The recent change in administration in Washington has contributed to a renewed sense of urgency around environmental issues,"* said Leahruth Jemilo, head of the ESG advisory practice at Corbin Advisors. Also, to mention that AI companies are supporting banks and financial institutions through software programs to assist in processing economic data. Other countries may follow the American model, to combat global warming. More data on this issue are found in paragraph 6.2 At the end of chapter 4, I present more examples of big private companies adjusting their policy to meet climate agreements, contribute to climate projects mitigation.

Chapter 5 In this chapter, I analyze the primary green sources of energy. Global fossil fuel share of primary energy: Oil 31.1%, Natural gas 24.2%, Coal 27%, Green energy 17.7%. Fossil fuel energy represents 82.3 %, versus renewable including nuclear and hydro 15.7 %, which is 5 times less than fossil fuel share. To reach net-zero emission by 2050, or earlier, is a very long, hard, and costly, achievement. Society needs to substitute 5 times more green energy than total uses. It is a very challenging task. The transition period should be as short as possible.

The estimated cost of reducing carbon emissions is falling rapidly. One dramatic example is an analysis by Geoffrey Heal of Columbia University showing that *"it would cost only $6 billion a year for the U.S. to move to carbon-free electricity generation by 2050."* (Note: This value is highly underestimated, other estimates say that the cost shall be in the trillions of dollars). *"Three forces are changing the math."*

First, renewable power costs are dropping so fast, both utility-scale *"solar and onshore wind power have become cheaper than natural gas or coal power."*

Second, the cost of storing renewable energy is also falling.

Third, and crucially, many power plants are nearing the end of their useful lives and need to be replaced one way or another. All these factors are good news for green energies. A short presentation of the most important aspects of new renewable solar and wind farms is presented. In detail are presented the hydrogen industry development on all parts of the planet.

It is unnormal that there is so little if any information about renewable energies in Russia, which has so far less than 1% of total energy. Nuclear energy has the potential to regain the market. This is due to new innovative technologies at the fission nuclear power plant (PP). An example is the first Arab new NPP in the United Arab Emirates. A remarkable development takes place in Small Modular Reactors (SMR). Also, to mention that a trillion dollars are invested in nuclear fusion technology, with leading countries: The USA, European Union, Britain, China, and South Korea. Japan and France are more inclined to build new nuclear power plants. If society wants to slow down global warming, then nuclear energy has to be included in the new actions.

A huge potential of developments have geothermal and wave energies. For instance wave energy has intensity per sq. meter, 10 times higher than solar energy. Geothermal energy is clean, dense, and countable. Iceland has the most developed geothermal system, 40% of the capital Reykjavik is heated with this energy. Reykjavik is one of the cleanest and unpolluted cities in the world

Chapter 6. The impact of climate change is spreading on each corner of the physical planet, and any angle of life on the Earth. From human life to biodiversity, ecosystems, ocean, ground, and air wildlife are disturbed, which will affect the safety and security of humanity and life on the planet. This all is interconnected on the planet, without considering borders, or regional limits of influences of governments or big industrial, and financial monopolies. The climate and weather phenomena don't have borders, same for birds, and fishes. The impact is global, from China to America, and from Russia to Brazil. Nature is declining globally at rates unprecedented in human history, and the *"rate of species extinctions is accelerating"*, with grave impacts on people around the world now likely, warns a landmark new report from the Intergovernmental Science-Policy Platform on Biodiversity and Ecosystem Services (IPBES). The most notable findings of the report presented is that: *"1 million animal and plant species are now threatened with extinction, many within decades, more than ever before in human history."*

Since 1970, vertebrate animals, birds, mammals, reptiles, fish, and amphibians, have declined by 60%. The world's last two northern white rhinos feature in the film done by Sir David Attenborough. These great beasts used to be found in their thousands in Central Africa but have been pushed to the brink of extinction by habitat loss and hunting. The list of extinction plants and animals is long. To mention some of them: whales, sharks, rays, millions of birds in the Pacific, and more. Arctic ice melting has reached the tipping point, and can not be stopped or slowed down. The warming of the Atlantic ocean is in the direction of reaching a tipping point too. These two phenomena started to increase the sea level. We need to mention here the catastrophic impacts of glaciers melting, like in the Himalayas and Alps mountain. Some of the impacts of global warming on humans, and presented in detail are the *"scarcity of food, water, and health services"*. Only due to pollution in 2019 more than 6.7 million people died. Population growth and migration are two factors analyzed, with scary consequences for humanity, which may lead to conflicts, even wars. Experts agree that a prolonged drought may have catalyzed Syria's civil war and resulting migration. *"Between 2008 and 2015, an average of 26.4 million people per year were displaced by climate- or weather-related disasters, according to the United Nations"*. And the science of climate change indicates that these trends are likely to get worse. Climate change will affect almost everyone on the planet to some degree, *"but poor people in developing nations will be affected most severely"*. Extreme weather events and tropical diseases wreak the heaviest damage in these regions. On top of all undernourished people who have few resources and inadequate housing are especially at risk and likely to be displaced.

Today the global community has not universally acknowledged the existence of climate migrants, much less agreed on how to define them. According to international refugee law, *"climate migrants are not legally considered refugees"*. Therefore, they have none of the protections officially accorded to refugees, who are technically defined as people fleeing persecution. No global agreements exist to help millions of people who are displaced by natural disasters every year. Rapid population growth, lack of access to food and water, and increased exposure to natural disasters mean *"more than 1 billion people face being displaced by 2050,"* according to a new analysis of global ecological threats. Compiled by the Institute for Economics and Peace (IEP), a think-tank that produces annual terrorism and peace indexes, the Ecological Threat Register uses data from the United Nations and other sources to *"assess eight ecological threats and predict which countries and regions are most at risk."* With the world's population forecast to rise to nearly 10 billion by 2050, intensifying the scramble for resources and fueling conflict, the research shows as many as *"1.2 billion people living in vulnerable areas of sub-Saharan Africa, Central Asia, and the Middle East"* may be forced to migrate by 2050. *"This will have huge social and political impacts, not just in the developing world, but also in the developed world, as mass displacement will lead to larger refugee flows to the most developed countries,"* said Steve Killelea, IEP's founder. Also, global warming is affecting the security of the countries, their arms bases, and nuclear plans with drastic impacts on society. The last paragraph of the book is Action. Is so difficult to incorporate all that is needed for slowing down global warming. First, one country or a group of countries can not do it. Has to be a global approach to the issue, as climate change is global, and does not have borders. Secondly, humanity has to act on what sources are producing the warming, and how the consumers of energy can reduce their consumption through innovation and breakthrough technologies. The emitters of GHG have to cooperate and collaborate for a universal plan. For instance, the United States and Chinese officials face pressure from nationalists at home to intensify what looks increasingly like a cold war between the two countries. Yet brokering a *"cooperative peace between the U.S. and China will be key to any global decarbonization plan"*. Abandoning plans to contain China's growth and instead of working together could help quickly deploy clean energy and climate solutions worldwide. To decarbonize the economy requires some global action, like carbon taxes, to note that the carbon taxes may very well be local (country) policy. With today's technology and increasing demand for fossil fuel, direct air capture technology needs to apply to decarbonize the atmosphere. We can not supersaturate, or add more GHG to the atmosphere. If the existing quantity produces such disastrous weather events, what to expect if we continue to add more GHG? Carbon prices or the way to get to net zero is a huge headache for all parties involved in decarbonization, from political to financial, industry, society, and individual citizens. And that complicates the road to net zero. In the paragraph are given many details, what comes out of this is that nothing can be done without sacrifice, mental - to accept it, morally- is it ethically, and financially- cost money. For instance, the IMF has estimated that household electric bills would rise 43% on average over the next decade if carbon was taxed appropriately. Rises would be steeper in countries that rely heavily on coal. In the U.S. major financial trade groups on February 18, 2021, joined calls for some type of carbon pricing, underscoring the growing corporate enthusiasm for steps to slow the pace of climate change.

A policy document of the group is set to release with 10 other trade groups that state that carbon pricing can *"spur development of climate-related financial products, promote more transparent pricing of climate-related financial risks, and can inform and help scale key initiatives like voluntary carbon markets."* I promote the idea of a Planet Tax, based on *the footprint of each citizen or entity on the planet Earth.* A carbon tax or price is a very sensitive issue for American politicians as well as for the Chinese government. Everything regarding this subject is under debate. For instance, the present 2020 carbon price to be between $40. and $80., and reach $100 by 2030. The big elephant in the backyard is net-zero, or carbon-neutral by 2050. The worldwide effort to prevent Earth from becoming an unlivable hothouse is in the grips of *"net-zero"* fever. *"More than 110 countries have committed to becoming carbon neutral by mid-century."* The European Union has taken the vow, as has US President Joe Biden. China, which *"generates 28% of all carbon pollution",* set 2060 as the year when any remaining emissions from energy, agriculture, or industry must be offset by tree farms or experimental technologies that suck CO_2 from the air. *"More than 65% of global CO_2 emitters now fall under such pledges".* The London-based Energy & Climate Intelligence Unit calculates the aggregate GDP of nations, cities, and states with *2050 net-zero targets is* *"$46 trillion, well over half of global GDP. I firmly believe that 2021 can be a new kind of leap year, the year of a quantum leap towards carbon neutrality,"* UN chief Antonio Guterres said in Dec. 2020 in New York. *"Every country, city, financial institution and company should adopt plans for transitioning to net zero emissions by 2050."* This is a big task, with no short-term details or commitments. The UN makes a very insistent push to governments to do more about climate change, to declare in their country a state of emergency for the climate crisis. Not only do governments through policy and regulations have to fight for global net-zero, but the big fossil fuel companies have to do it. There are lots of details about the way big fossil and financial companies try to hide their businesses, showing compliance with net-zero. A special paragraph is dedicated to the global elite (rich) high carbon lifestyle. Education and denial of the climate crisis, details the 5 categories of denial, to make the public believe that the climate crisis does not exist. The denials are science, economic, humanitarian, political, and crisis

ANTHROPOCENE

1.1. THE STARTING TIME IS 1610

In the last 2 millennia of human civilization, **Technology and Science (TS)** have not developed in an even linear ascending way, not to mention an exponential way. So the cold Black Eve, which lasted for more than 15 centuries, was shadowed by Christian religion MOTO of *"Belief and do not Search"*. This dogma of the church and the inquisition has impeded the TS of an ascendant development. *"One can only speculate where our civilization might be today if this ancient science had continued unabated"*. Even Galileo was scared of the drastic measure taken by the church inquisition against those who do not follow their dogmas. Newfound Galileo's letter suggests he lied to dupe the church and avoid persecution.

Aside from his many contributions to the evolution of modern science, Galileo Galilei is also famous for defying the 17th century Catholic Church. He's lionized for advocating a heliocentric view of our solar system, spurning Church doctrine, and becoming a heretic. But a discovery presents an interesting wrinkle in the Galileo narrative. In the same period of time, Giordano Bruno was burned for his beliefs in the heliocentric solar system (7).

At the same time, geologists and environmentalists discovered that Planet Earth began to experience the first signs of **CLIMATE CHANGE.** Today our impacts on the environment are immense: humans move more soil, rock, and sediment each year than is transported by all other natural processes combined. We may have kicked off the sixth *"mass extinction"* in Earth's history, and the global climate is warming so fast that we might have delayed the next ice age.

We've made enough concrete to cover the entire surface of the Earth in a layer two millimeters thick. Enough plastic has been manufactured to clingfilm it as well. We annually produce 4.8 billion tonnes of our top five crops and 4.8 billion livestock animals. There are 1.4 billion motor vehicles, 2 billion+ personal computers, and more mobile phones than the 7.8 billion people on Earth. All this suggests humans have become a *"geological superpower"* and evidence of our impact will be visible in rocks millions of years from now. This is a new geological epoch that scientists are calling the **ANTHROPOCENE**, combining the words, *"human"* and *"recent-time"*. But debate continues as to when we should define the beginning of this period. When exactly did we leave behind the Holocene, the *"10,000 years of stability"* that allowed farming and complex civilizations to develop, and moved into the new epoch? Evidence shows that *"the start of capitalism and European colonization meet the formal scientific criteria for the start of the Anthropocene"*. Our planetary impacts have increased since our earliest ancestors stepped down from the trees, at first by hunting some animal species. Much later, following the development of farming and agricultural societies, we started to change the climate. Earth only became a *"human planet"* with the incipient roots of a new society. This was capitalism, which itself grew out of European expansion, and the era of colonization of indigenous peoples all around the world. In the Americas, just 100 years after Christopher Columbus first set foot on the Bahamas in 1492. Diseases brought by European killed more than 56 million indigenous Americans, mainly in South and Central America. This was 90% of the population. Indigenous populations had never experienced diseases like measles, smallpox, influenza, the bubonic plague. War, slavery, and disease combined to cause this **"Great Dying"**, something the world had never seen before, or since. In North America, the population decline was slower but no less dramatic due to the slower colonization by Europeans. US census data suggest the Native American population may have been as low as 250,000 people by 1900 from a pre-Columbus level of 5 million, a 95% decline. Most pre-Columbian indigenous Americans were farmers. After their death, previously managed landscapes returned to their natural states, absorbing more carbon from the atmosphere. The drop in atmospheric carbon dioxide recorded in Antarctic ice cores was around the year **1610.** The reconnection of the continents and ocean basins for the first time in 200 million years has set Earth on a new developmental trajectory. The drop in carbon dioxide at 1610 provides a first marker in geological sediment which provides a *"sensible start date for the new Anthropocene epoch" (8)*. If 1610 marks both a turning point in human relations with the Earth and our treatment of each other, *"then 1859 represents the year of the birth of Climate Science"*, thus he is considered the father and co-discoverer of it, John Tyndall.

It is surprising that the Irish scientist John Tyndall, born 201 years ago on August 2, 1820, is not better known. In 1859, Tyndall showed that gases including *"carbon dioxide and water vapor can absorb heat."* His heat source was not the Sun, but radiation from a copper cube containing boiling water. In modern terms, this was infrared radiation, just like that emanating from the Earth's surface. Previous work had shown that the Earth's temperature was higher than expected, which was put down to the atmosphere acting as an insulator. But no one knew the explanation for what we now call the greenhouse effect, gases in the atmosphere trapping heat. What Tyndall did was discover and explain this mechanism.

He wrote: *"Thus the atmosphere admits of the entrance of the solar heat but checks its exit, and the result is a tendency to accumulate heat at the surface of the planet."* He realized that any change in the amount of water vapor or carbon dioxide in the atmosphere could change the climate. His work, therefore, set a foundation for our understanding of climate change and meteorology. American Eunice Foote, who showed in 1856 using sunlight that carbon dioxide could absorb heat. She suggested that an increase in carbon dioxide would result in a warmer planet. (John Tyndall. Wellcome/ Wikipedia, CC BY-SA Research suggests).

He worked on the absorption of heat by gases, and then the action of light in causing chemical changes. In the process *"Tyndall explained why the sky is blue"*: blue light is scattered more by gases in the sky than other colors because of its short wavelength. Tyndall made many other discoveries in disparate fields of physics and biology. Within a few years, he was a fellow of the Royal Society, Britain's most prestigious scientific body, and professor of natural philosophy at the Royal Institution, where he remained for the rest of his scientific career. Soon he was at work on understanding glacier structure and motion. He also discovered *"Tyndallization, a bacteriological technique of sterilization when undertaking experiments alongside French biologist Louis Pasteur to support the theory that germs can cause disease.* Tyndall would be gratified to find that his foundational work had proved so important. Climate science is now the future rather than the past, and it is, therefore, time to recognize and reinstate Tyndall as a major Irish scientist, mountaineer, and public intellectual (8a).

Coming back to the present, ***"2020**** could mark the start of the fight on Climate Crisis, environmental, (species and plants extinction), humanitarian justice and stewardship of the only planet in the universe known to harbor any life"*. It's a struggle nobody can afford to lose.

"The 2020 year" is not only a breaking year for the climate crisis, *"is a galactic year for the Planet Earth, is the year when a **Health Crisis overlaps Climate Crises. Humanity has experienced more Pandemics in its history, but never one with so many implications in the life of humanity". Practically the Pandemic spread through the entire planet and humanity.*** Here I wish to present only one aspect of **Covid-19** implication in the Science of the Planet. COVID-19 lockdowns worldwide led to the longest and most pronounced reduction in human-linked seismic vibrations ever recorded, sharpening scientists' ability to *"hear earth's natural signals and detect earthquakes"*, a study found on Thursday, July 23, 2020. Vibrations travel through the earth like waves, creating seismic noise from earthquakes, volcanoes, wind, and rivers as well as *"human actions"* such as travel and industry. In the study, published in the journal Science and conducted using international seismometer networks, scientists found that human-linked earth *vibrations dropped by an average of 50% between March and May this year, 2020. The 2020 seismic noise quiet period is the longest and most prominent global anthropogenic seismic noise reduction on record."* The work was co-led by the Royal Observatory of Belgium and five other institutions using data from 268 monitoring stations in 117 countries. Beginning in China in late January, and followed by Europe, researchers saw *"a wave of quietening"* as a worldwide lockdown. Travel and tourism were all but halted, millions of schools and industries closed, and many people were confined to their homes. The relative quiet allowed

scientists to *"listen in"* in more detail on the earth's natural vibrations. *"It has yielded a new window on the natural seismic signals, and could let us see more clearly than ever what differentiates human and natural noise."* *(9).*

1.2. INDUSTRIAL ERA

Starting with technology and development, the innovation of the steam engine and electricity in the 18th and 19th centuries led to the beginning of the industrial era of humanity. Steam engines lead to the use of coal. The Discovery of Electricity by Thomas Edison has been a breakthrough technology on its own, which revolutionized dozens of different industries like lighting buildings and cities, cooking foods, empowering industrial and agricultural equipment, transportation, and more. The same coal and oil became the major fuels for producing electricity. In that time the use of fossil fuels has intensified for producing electricity and use in combustion engines not only in steam engines. These types of developments have taken millions of farmer workers from fields to factories and developed industrial zones close to cities and lately to megacities. After the Second World War technology and science have taken an exponential ascent in a few most developed countries in the world, specifically in the USA, Europe, Japan, Canada, and the Soviet Union, today Russia. In 1957 the Soviet Union launched the first human-made satellite in orbit. This achievement by the Soviets had a profound effect on the American Government, which led to the creation of the *"National Astronautics and Space Administration"* *(NASA).* American Government poured billions of dollars into math and science education and started the space program. Twelve years later **Neil Armstrong** was the first-ever person **to put his foot on the Moon.** But these waves of TS developments have created the best conditions for advanced computing in all aspects of science, which leads to the appearance of **Information Technology (IT).** The need for computing data in the highest aspects of science and technology gave birth to *"Computing Science"* and *"Computing Machines"* (Computers) companies, specifically in the US, Canada, Japan, and other countries in Europe. The most known area of this development has been in Silicon Valley in the USA, the area between San Francisco and San Jose cities (as known as Bay Area). This book will not present a history of computer and computer science developments that lead to machine learning to do programming, known as **Artificial Intelligence (AI)**. In his book, *"AI Super Powers, China, Silicon Valley, and the New World Order"* Mr. Kai - Fu Lee does a very detailed presentation of the History and the Development of IT, AI, and Super AI (SAI) in the USA and China, from earlier 1950 until the end of 2018 when the book was published (10). In the area of computing, one of the first pioneering companies in the USA was the *"International Business Machine (IBM) and Microsoft"*. To show one of the most interesting achievements of this company, which compares human intelligence versus machine intelligence, is the *"IBM Deep Blue Computer"* which defeated world chess champion, Garry Kasparov, in a 1997 match called *"The Brain's Last Stand"*. That event put the question *'When the machine/robot would launch their conquest of humankind'?* That has been the start of *"Deep Learning or Neural networks"*, which takes a different approach in rules of computing. Instead of teaching the computers the rules mastered by the human brain, this neural network tries to reconstruct the human brain itself. One of the researchers who developed an effective way to train the layer of neural networks was Geoffrey Hinton in the mid-2000s and constitutes the beginning of AI.

Lord Martin Rees in his book "On the Future, Prospective for Humanity," presents some interesting examples, like Indians, who now have an electronic identity card that makes it easier for them to register for welfare benefits. This card doesn't need a password. The vein patterns in our eyes allow the use of *"iris recognition"* software, a substantial improvement from fingerprints or facial recognition. Employers can control their workers far more intrusively, which leads to more privacy concerns. *"Are you happy if a random stranger sitting near you in a restaurant or on public transportation can, via facial recognition, identify you, and invade your privacy?"(11).*

Singularity, or (ASI). These machines will improve and develop so fast, that they will have the ability to perform any intellectual task that a human can not, and much more. This is considered **The Holy Grail** of AI research or **Artificial General Intelligence (AGI).** *"When machines surpass human intelligence and can automatically program themselves"*, they achieved the so-called **Artificial Superintelligence** or **Singularity** (10). Ray Kurzweil -the eccentric inventor, futurist, and guru-in-residence at

Google, predicts that by 2029 we will have computers with intelligence comparable to that of humans and that we will reach *"Singularity"* by 2045 (10).

For more about Singularity and its achievements, see *"Humanity Must Survive the 21ˢᵗ Century"* by Prof. Theodore Vornicu (6).

Some utopists are positive thinking that SAI will be helpful for humanity. The computer may develop *"minds of their own"* and pursue goals hostile to humanity. Would a powerful futuristic SAI remain docile, or *"go rogue"*(11). Some are not so optimistic (dystopian) like Elon Musk who has called Superintelligence *"the biggest risk we face as a civilization, comparing the creation of it to "summoning the demon".* The famous physicist, cosmetologist, and mathematician **Stephen Hawking** has joined Musk in the dystopian camp. Stephen Hawking predicted A.I. maybe *"The Worst Thing"* for humans. In November 2017, Hawking warned, at a technology conference in Lisbon, Portugal, that artificial intelligence could be *"the worst thing ever to happen to humanity."* Because what an A.I. can learn is infinite, Hawking reasoned that it could eventually catch up to the limits of the human brain and surpass us. *"Success in creating effective A.I. could be the biggest event in the history of our civilization or the worst,"* Hawking said at Web Summit in 2017. *"We cannot know if we will be infinitely helped by A.I. or ignored by it and sidelined or conceivably destroyed by it."* Hawking also told *Wired* in November 2017, that he feared A.I. would *"replace humans altogether,"* a concern he had in common with Elon Musk. Accordingly, the two men, in February 2017, endorsed a list of 23 principles they feel should steer A.I. development. *"If we can augment our brain with electronic implants, we may be able to download our thoughts and memories into the machine. If these technical trends proceed unimpeded, then some people now living could attain immortality"* (11).

In the next 15 years, society will experience a technological disruption never seen before in the history of humanity. In the sectors of information, energy, transportation, food, and materials, this disruption will have huge transformations all over the economy. Technological innovations across these five sectors will drop production costs at least 10 times, putting industries from fossil fuels to dairy farming out of business. New discoveries and high forms of progress, the existing system gets to its limits increasing global crises, from climate crisis to pandemics, and more. Smartphones and the internet, the cloud and AI, disrupt telecommunications, entertainment monoliths, and taxi markets. These will lead to a new based system that would create the abundant, local and cheap production of energy, food, knowledge, tools, and materials for the benefit of all, rather than a few. That will drive the society to be free of concerns around material survival, the "Age of Freedom." Big sources of GHG will be unrecognizable before 2035, as big consumers of fossil fuels, and manufacturing and mining will diminish. Societies that will not adapt to these transformations will remain stuck in the dark ages unprepared for business failures and job losses, and incapable of reaping the benefits. Due to intelligent approaches, the disruption will lead to chaos and discontent. To avoid economical, political, and ecological instability and collapse of societies, humanity must understand the speed, scale, and dynamics of the new disruptions. The challenge humanity is facing needs complex systems thinking to solve. We need new models of thinking which engage with complex systems, relinquish disconnection for interconnection, and harness planetary collective consciousness for local implementation (8). Society to manage this emerging reality, needs a new mental approach, with awareness and intelligence of Nature events itself. All these transformations we need to engage our consciousness, where all the systems and activities of humanity and the natural world are integrated, harmonized, and mutually supportive. *"In this type of society trumpism and his adherents can not find a place. We can make a society full of unlimited possibilities, and it is up to us to make it if the Climate Crisis will not stop the US"*. We shall see how the entire society will treat the climate emergency in the next 3 decades, with an emphasis in the first 15 years from now (2021).

Science and Technology have taken an exponential development, especially in the last 2 decades (2000 to 2020). All these technologies are consuming enormous quantities of energy produced from fossil fuels: oil (petroleum products), natural gas, and coal. Also, energy is produced from nuclear and hydroelectric power plants, green sources like wind, solar, geothermal, wave, and biogases. Extracting and burning fossil fuels to produce energy, the outcome of these processes, besides energy, is the emission of *"greenhouse gases"* (GHG). The most known greenhouse gases are carbon dioxide CO2 and methane CH4. There are other gases like nitrogen oxides NOx and sulfur dioxide (SO2) that contribute to polluting the air we breathe. *"Emissions from cattle and other ruminants are almost as large as those from the fossil fuel industry for methane,"* Stanford University scientist Rob Jackson said in a statement. *"People joke about burping cows without realizing how big the source is"*. Methane is also the main ingredient in natural gas and is the second-largest contributor to global warming, after carbon dioxide. *"Methane, odorless and invisible, captures 86 times more heat than CO$_2$ over the first two decades of release, and at least 25 times more over a century."* More than half of all methane emissions now come from human activities. Annual methane emissions are up 9%, or 50 million tons per year, from the early 2000s when methane concentrations in the atmosphere were relatively stable. *"For instance in 2019, it will emit around* **39 Gigatons** *of greenhouse gases"*. (This value varies between 37 and 42 gigatons, depending on the sources). Note: 1Gigaton = 1 billion tons. If carbon dioxide has been lingering in the atmosphere for more than 100 years, how much carbon do we have today in the atmosphere? The answer is approximately 420 Gigatons. Before presenting: How global warming is developed? Let see some simple notions encountered in Climate Change per Wikipedia. (14).

What is Climate Change?

All these definitions and descriptions are from Atmospheric Science and Climatology University Courses, as well as from National Oceanic and Atmospheric Administration. The Earth's average temperature is about 15 C but has been much higher and lower in the past. There are natural fluctuations in the climate but scientists say temperatures are now rising faster than at many other times. This is linked to the greenhouse effect, which describes how the Earth's atmosphere traps some of the Sun's energy. Solar energy radiating back to space from the Earth's surface is absorbed by greenhouse gases and re-emitted in all directions.

This heats both the lower atmosphere and the surface of the planet. Without this effect, the Earth would be about 30 C colder and more hostile to life.

Scientists believe we are adding to the natural greenhouse effect, with gases released from industry and agriculture trapping more energy and increasing the temperature.

"This is known as climate change (CC) or global warming (GW)". (Wikipedia)

What Are Greenhouse Gases?

The greenhouse gas with the greatest impact on warming is water vapor. But it remains in the atmosphere for only a few days. Carbon dioxide (CO_2), however, persists for much longer. It would take hundreds of years for a return to pre-industrial levels and only so much can be soaked up by natural reservoirs such as the oceans, and large forestry, like tropical forests (e.a. Amazon forest). Most man-made emissions of CO_2 come from burning fossil fuels. When carbon-absorbing forests are cut down and left to rot, or burned, that stored carbon is released, contributing to global warming. One more intensive greenhouse gas is methane, which has an 89 times stronger effect on GW than CO_2 in the first 20 to 30 years. Since the Industrial Revolution began in about 1750, CO_2 levels have risen more than 30%. *"The concentration of CO_2 in the atmosphere is higher than at any time in at least 800,000 years."* (12). Other greenhouse gas pollutants are nitrogen oxides (NOx), sulfur dioxide SO_2, and mercury (14).

What is the Evidence For Warming?

The world is about 1.2 degrees Celsius warmer than before widespread industrialization, according to the World Meteorological Organization (WMO). It says the past five years, *"2015–2019, were the warmest on record."* Across the globe, the average sea level increased by 3.6 mm (0.142 inches) per year between 2005 and 2015. Most of this change was because water increases in volume as it heats up. However, melting ice is now thought to be the main reason for rising sea levels. Most glaciers in temperate regions of the world are retreating. And satellite records show a dramatic decline in Arctic sea ice since 1979. The Greenland Ice Sheet has experienced record melting in recent years. Satellite data also shows the West Antarctic Ice Sheet is losing mass. A recent study indicated East Antarctica may also have started to lose mass. The effects of a changing climate can also be seen in vegetation and land animals. These include earlier flowering and fruiting times for plants and changes in the territories of animals (14).

How Much Will Temperatures Rise in The Future?

The change in the global surface temperature between 1850 and the end of the 21st Century is likely to exceed 1.5C, most simulations suggest. The WMO says that if the current warming trend continues, temperatures could rise 3-5C (5.4-6.5F) by the end of this century.

Scientists and policymakers have argued that limiting temperature rises to 1.5C is safer.

The Intergovernmental Panel on Climate Change (IPCC) report in 2018 suggested that keeping to the *1.5C target would require rapid, far-reaching and unprecedented changes in all aspects of society"* (14).

How Will Climate Change Affect Us?

There is uncertainty about how great the impact of a changing climate will be, but as the weather showed last period of time (Sept. 2019 to April 2021), the impact will have *"catastrophic effects"*. It could cause freshwater shortages, dramatically alter our ability to produce food, and increase the number of deaths from floods, storms, and heatwaves. Also, huge population migration and conflicts, which may even go to *"military wars that might be the day norms"*. This is because climate change is expected to increase the frequency, and intensity of extreme weather events. As the world warms, more water evaporates, leading to more moisture in the air. This means many areas will experience more intense rainfall - and in some places snowfall. But the risk of drought in inland areas during hot summers will increase. More flooding is expected from storms and rising sea levels. But there are likely to be very strong regional variations in these patterns. Poorer countries, which are least equipped to deal with rapid change, could suffer the most. Plant and animal extinctions are predicted as habitats change faster than species can adapt. And the World Health Organization (WHO) has warned that the health of millions could be threatened by increases in malaria, a water-borne disease, malnutrition, and pollution. As more CO2 is released into the atmosphere, uptake of the gas by the oceans increases, causing the water to become more acidic. This, and the increased temperature of the ocean surface layer of water, could pose major problems for coral reefs, like bleaching and death. Global warming will cause further changes that are likely to create further heating. This includes the release of large quantities of methane as permafrost, and peatlands, frozen soil found mainly at high latitudes (Siberia, Greenland, and Alaska), melts. *Responding to climate change will be the biggest challenges we face this century."* (Wikipedia).

How Global Warming is Developed?

The Sun's beams penetrate the atmosphere and reach the ground, heating soil, and water. The Sun temperature is about 2 million degrees Celsius, and the beames emitted by the Sun have a very short length wave, so they can travel (penetrate) through the Solar system. On reaching the atmosphere, the Sun's beams also heat it. Part of the Sunbeams reaching the ground is reflected in the atmosphere. The atmosphere and all the ground of the planet are heated at a temperature much much smaller than the emitter. The reflected beam's length is much higher and can not reach far beyond the lower atmosphere. And, here is the **Catch,** the greenhouse gases in the atmosphere (mainly CO2, CH4 ....) are blocking the heat to leave the atmosphere, *"so everything in the lower atmosphere and ground is heating up."*

Humanity through the release of greenhouse gases is changing the equilibrium of the Ecological, Environmental, and Atmospheric physical conditions of Planet Earth.

Do we know how these changed conditions on the planet are acting against all physical phenomena on Earth? Not really, but we are witnessing an increase in the intensity and frequency of all Climatic Phenomena. Since the middle of 20-century scientists has raised the issue of climate change and that the planet is warming. The consequences of planet-warming started to raise the eyebrows of politicians, governments, businesses, as well as scientists from different fields of activities like environmental, ecological, meteorological, cosmological, and so on, including NASA scientists. Almost all the parties present the catastrophic consequences of global warming and make recommendations for reducing the causes of climate change. The bottom line is reducing the emission of greenhouse gases, deforestation, changing the way agriculture is done nowadays, and people's lifestyle. For these to be achieved, humanity needs a Global Agreement achieved and issued under the umbrella of the **United Nations** (Wikipedia).

1.3. THE PARIS AGREEMENT

Before I get into detailing the Paris agreement, I wish to make an introduction showing the

"dirty tricks climate scientists have faced in the three decades since the first report by IPCC was released." Thirty years ago, in a small Swedish city called Sundsvall, the Intergovernmental Panel on Climate Change (IPCC) released its first major report. From that time scientists have fought the well-paid lobbyists of big fossil fuel companies. The geologist and climate campaigner

Jeremy Leggett has storied the encounter with Don Pearlman, a lobbyist for coal business. The representative of the Pacific island of Kiribati expressed concerns about global warming for low-lying islands. He concluded: "I hope this meeting will not fail us. Thank you. Powerful and rich countries in the world, Russia, Saudi, and American's reps cut away the sensitive and alarming words, being unclear and uncertain. The GHG has worn the scientist since earlier 1960-1970. Scientists gain respect and credibility among the public, after the discovery of an ozone hole above Antarctica. The IPCC banned the chemical chlorofluorocarbons, the substance causing the issue. In the USA fossil fuels industries set up the Global Climate Coalition to oppose action on climate change and doubted the evidence, before the Sundsvall IPCC meeting in 1989. The scientist insisted, and in 1992 a treaty was signed. Attempts to discredit the scientist did not stop. In the 1990s, the hockey stick diagram by Michael Mann showing the rise of global temperature came as a pledge for fossil fuel companies. *"As the evidence became ever more compelling, the attacks on scientists escalated"* The CC deniers hacked scientist email before the Copenhagen summit. They tried to show the scientists are guilty of *"scaremongering"*. The deniers fought hard to create doubts over the scientific proofs and consensus (UN - IPCC). This fight for scientists against personal financial interests is going on even today. The warming of the atmosphere, land, and oceans, is not a local phenomenon, is ***"global and does not have boundaries"***. [The effect of Amazon deforestation has an impact on more than 3000 miles (4800 km.) away in California's weather, based on Berkeley Univ. study of Climate Change]. Any effort to reduce the increase of greenhouse gases and control the industry that emits them should be global. Therefore the fifth report from the Intergovernmental Panel on Climate Change (IPCC), published in 2013, clearly underlines that *"increase mainly of CO2 and CH4 is not checked and regulated,"* we risk drastic climate change with the devastating increase of ocean water level, intensifier floodings, tornados, hurricanes, typhoons, drought, fire, environmental destruction and species of plants and animal extinction, for decades ahead. And most importantly, if the *"tipping points of climate change are reached, then the increase of intensity and frequency of most weather phenomena become irreversible"*

The **Paris Agreement** (French: *l'accord de Paris*) is an agreement within the United Nations Framework Convention on Climate Change (UNFCCC), dealing with greenhouse-gas-emissions mitigation, adaptation, and finance, signed in 2016. The agreement's language was negotiated by representatives of 196 state parties at the 21st Conference of the Parties of the UNFCCC in Le Bourget, near Paris, France, and adopted by consensus on 12 December 2015. As of February 2020, all UNFCCC members have signed the agreement, 189 have become a party to it, and the only significant emitters which are not parties are Iran and Turkey. The Paris Agreement's long-term temperature goal is to keep the increase in global average temperature to well below 2°C above pre-industrial levels and to pursue efforts to limit the increase to 1.5°C. This should be done by reducing emissions as soon as possible, to *"achieve a balance between anthropogenic emissions by sources and removals by sinks of greenhouse gases" in the second half of the 21st century. It also aims to increase the ability of parties to adapt to the adverse impacts of climate change, and make "finance flows consistent with a pathway towards low greenhouse gas emissions and climate-resilient development."* Under the Paris Agreement, each country must determine, plan, and regularly report on the contribution that it undertakes to mitigate global warming. In June 2017, U.S. President Donald Trump announced his intention to withdraw the United States from the agreement. Under the agreement, the earliest effective date of withdrawal for the U.S. is November 2020, shortly before the end of President Trump's 2016 term. In practice, changes in the United States policies that are contrary to the Paris Agreement have already been put in place (Per United Nations released public document). (Wikipedia)

Aims of The Agreement

Decrease global warming, *"enhancing the implementation"* of the UNFCCC through:

(a) Holding the increase in the global average temperature to well below 2 °C above pre-industrial levels and to pursue efforts to limit the temperature increase to 1.5 °C above pre-industrial levels.

(b) Increasing the ability to adapt to the adverse impacts of climate change and foster climate resilience and low greenhouse gas emissions development, in a manner that does not threaten food production; (c) Making finance flows consistent with a pathway towards low greenhouse gas emissions (later agree to net-zero emission of greenhouse gases by 2050).and climate-resilient development. The agreement has been described as an incentive for and driver of fossil fuel divestment. **The Paris deal is the world's first comprehensive climate agreement.**

Global Temperature

The negotiators of the agreement, however, stated that the Nationally Determined Contribution (NDCs) and the target of no more than 2°C increase were insufficient; instead, *"a target of 1.5°C maximum increase"* by 2100. **Author Note:** This value will be reached before 2026. When the agreement achieved enough signatures to cross the threshold on 5 October 2016, **US President Barack Obama** claimed that *"Even if we meet every target ... we will only get to the part of where we need to go."* He also said that *"this agreement will help delay or avoid some of the worst consequences of climate change. It will help other nations ratchet down their emissions over time, and set bolder targets as technology advances, all under a strong system of transparency that allows each nation to evaluate the progress of all other nations."* (14).

New Climate Models Suggest Paris Goals May Be Out Of Reach

New climate models show carbon dioxide, CO_2, is a more potent greenhouse gas than previously understood, *"a finding that could push the Paris treaty goals for capping global warming out of reach"*, scientists have told AFP. (**Note**: As the events have developed since 2016, the UN targets on climate change become obsolete, and out of reach. The content of the book and last chapter *"Conclusion"* will prove these statements). The new models suggest *scientists have for decades consistently underestimated the warming potential of CO_2."*

We have better models now, Olivier Boucher, head of the Institut Pierre Simon Laplace Climate Modelling Centre in Paris, told AFP, adding that they *"represent current climate trends more accurately"*. The most influential projections from government-backed teams in the US, Britain, France, and Canada point to a future in which CO_2 concentrations that have long been equated with a 3C world would more likely heat the planet's surface by 4C or 5C degrees.

"If you think the new models give a more realistic picture, then it will, of course, be harder to achieve the Paris targets, whether it is 1.5 or 2 degrees Celsius," scientist Mark Zelinka told AFP. Zelinka, from the Lawrence

Livermore National Laboratory in California, is the lead author of the first peer-reviewed assessment of the new generation of models.

'Holy Grail' For Scientists

For more than a century, scientists have puzzled over a deceptively simple question: if the amount of CO_2 in the atmosphere doubles, how much will Earth's surface warm over time?

The resulting temperature increase is known as Earth's "climate sensitivity".

That number has been hard to pin down due to a host of elusive variables. We assume that oceans and forests, for example, will continue to absorb more than half of the CO_2 emitted by humanity, but it is hard to predict. *"How clouds evolve in a warmer climate and whether they will exert a tempering or amplifying effect has long been a major source of uncertainty,"* explained Imperial College London researcher Joeri Rogelj, the lead IPCC author on the global carbon budget, the number of greenhouse gases that can be emitted without exceeding a given temperature cap. These days the concentration stands at 412 ppm, a 45% raise, half of it in the last three decades. (New Delhi, India, reached in Nov. 2020, 425. ppm).

Last year (2019) human activity injected more than 39 billion tonnes of CO_2 into the atmosphere, some five million tonnes per hour. (**Note:** The existing GHG in the atmosphere is estimated at 420Gigatones). By the late 1970s, scientists settled on a climate sensitivity of 3C (plus-or-minus 1.5C), corresponding to about 560 ppm of CO_2 in the atmosphere. (**Note:** At 560 ppm in the air no human can breathe). That assessment remained largely unchanged, until now. The IPCC, the UN's climate advisory body, posits four scenarios for future warming, depending on how aggressively humanity works to reduce greenhouse gases.

-The most ambitious, in line with the Paris goal of capping temperature rise to "well below" 2C, would require slashing CO_2 emissions by more than 10% per year, starting now, 2020. If we do the math, in 5 years the human injected CO_2 will be 23,64 billion tonnes (almost half of today's emission), and in ten years will be 14 billion tonnes of CO_2, a reduction of 66%.

-At the other extreme, a so-called "business-as-usual" trajectory of increased fossil fuel use would leave large swathes of the planet uninhabitable before the end of the century. (U.N. IPCC 2014 Report).

The first scenario has become wishful thinking, according to many scientists, while the *"worst-case is unlikely unless Earth itself begins releasing natural stores of greenhouse gases from, say, melting permafrost. That is scary." That leaves two middle-of-the-road scenarios, known as RCP4.5 and RCP6.0, that more likely reflect our climate future"*. According to the IPCC, *"the first would correspond to 538 ppm of CO2 in the atmosphere, while an RCP6.0 pathway would see an increase in CO2 concentration to 670 ppm."* Either way, the situation is very dangerous for human life. With only one degree Celsius of warming so far, the world is coping with increasingly deadly heatwaves, droughts, floods, and tropical cyclones made more destructive by rising seas *"You have 12 or 13 models showing sensitivity which is no longer 3C, but rather 5C or 6C with a doubling of CO2." "What is particularly worrying is that these are not the outliers."* Models from France, the US Department of Energy, Britain's Met Office, and Canada show climate sensitivity of 4.9C, 5.3C, 5.5C, and 5.6C respectively, Zelinka said.

"You have to take these models seriously, they are highly developed, state-of-the-art."

"Climate sensitivity has been in the range of 1.5C to 4.5C for more than 30 years. If it is now moving to between 3C and 7C, that would be tremendously dangerous."

Global Carbon Dioxide Emissions by Jurisdiction, The Year 2016.

Here are the most emitters of CO2: China (29.4%), United States (14.3%), European Economic Area (9.8%), India (6.8%), Russia (4.9%), Saudi Arabia (4.70%), Japan (3.5%), Others (31.3%).

During the Katowice Climate Change Conference in Poland in December 2018, to contain warming at 1.5C, man-made global net carbon dioxide (CO2) and methane (CH4) emissions would need to fall by about *"45% by 2030"* from 2010 levels and reach *"net-zero by mid-century.* Any additional emissions would require removing CO2 from the air." *The report shows that we only have the slimmest of opportunities remaining to avoid unthinkable damage to the climate system that supports life as we know it,"* said Amjad Abdulla, the IPCC board member and chief negotiator for an alliance of small island states at risk of flooding as sea levels rise. (BP).

Grim Tidings From Science on Climate Change. Here are a few key points mentioned in the South Korea report. **1 degree Celsus.** Earth's average surface temperature from January to October 2018 was one degree Celsius (1.8 degrees Fahrenheit) higher than the 1850-1900 baseline. Long-term warming is caused by the accumulation of heat-trapping greenhouse gases in the atmosphere, especially carbon dioxide (CO_2) and methane (CH_4) cast off when fossil fuels are burned to produce energy. **405.5 ppm.** The concentration of carbon dioxide (CO_2) in the atmosphere reached 405.5 parts per million (ppm) in 2017, the highest in at least three million years and a 45% jump since the preindustrial era (280ppm). The last time CO_2 was at that level, oceans were 10-20 meters higher. Concentrations of the second most important greenhouse gas, methane (CH_4), have also risen sharply due to leakage from the gas industry's fracking boom and flatulence from expanding livestock.

Emissions. After remaining stable for three years, carbon pollution increased more than one percent in 2017 to 53.5 billion tonnes of CO_2-equivalent, a measure that includes all main greenhouse gases. At that pace, Earth will pass the 1.5C marker as early as 2030. To cap global warming at 2C, emissions must stay under 1.5C, *"emissions will have to drop by more than half, and reach net-zero emission by 2050",* per the 2015 Paris Agreement. **Melting Ice.** Arctic summer sea ice shrank in 2018 to a low of 4.59 million square kilometers (1.77 million square miles), well above the record low of 3.39 million square kilometers set in 2012. But long-term trends are unmistakable: Arctic sea ice cover is declining at a rate of more than 29% per decade, relative to the 1981-2010 average. Climate models predict the Arctic Ocean could, in some years, be ice-free as early as 2030. **Extreme events.** The World Meteorological Organization (WMO) says there are clear links between climate change and increases in the intensity and frequency of extreme weather. The number of climate-related extreme events -- such as droughts, wildfires, heatwaves, floods, and cyclones -- has doubled since 1990, research has shown. The intensity of typhoons battering China, Taiwan, Japan, and the Korean Peninsula since 1980 has increased by 12% to 15%. *"Natural disasters drive more than 25* *million people into poverty every year"*, according to the World Bank, and cause annual losses above half a trillion dollars (440 billion euros). **84.8 millimeters.** Water that expands as it warms and runoff from ice sheets atop Greenland and Antarctica currently adds about three millimeters (0.12 inches) to sea levels per year. Since 1993, the global ocean watermark has gone up by more than 85mm (3.3 inches). That pace is likely to pick up, threatening the homes and livelihoods of tens of millions of people in low-lying areas around the world. Melting glaciers could lift sea levels at least 3 meters (9.5 feet) by 2100, and with the only 2C of warming by several meters more over the following centuries. (14) **1/5 of species are affected.** Of the 8,688 animal and plant species listed as *"threatened"* on the International Union for the Conservation of Nature's (IUCN) Red List, a fifth have been hit by climate change. From 1970 to 2014, the global population of vertebrates, birds, reptiles, amphibians, mammals, and fish plummeted by about 60 percent, due mainly to killing for food or profit, and habitat loss. The number of species is declining 100 to 1,000 times faster than only centuries ago, which means the planet has entered a *"mass extinction event"* only the sixth in the last half-billion years. Sources: NASA, NSIDC, UNEP, WMO, IPCC, NOAA, all public publications. (14). Trump last year 2017 announced his intention to withdraw the United States from the 2015 Paris Agreement *(How much ignorance and stupidity may be in the 21ˢᵗ century)*. The deal was agreed by nearly 200 nations to combat climate change, arguing the accord would hurt the U.S. economy and provide a little tangible environmental benefit. Trump and several members of his cabinet have also repeatedly cast doubt on the science of climate change, arguing that the causes and impacts are not yet settled. Environmental groups said the report reinforced their calls for the United States to take action on climate change. *"We are already seeing firsthand: climate change is real, it's happening here, and it's happening now."* Previous research, including from U.S. government scientists, has also concluded that climate change could have severe economic consequences, including damage to infrastructure, water supplies, and agriculture. Severe weather and other impacts also increase the risk of disease transmission, decrease air quality, and can increase mental health problems, among other effects. Thirteen government departments and agencies, from the Agriculture

Department to NASA, were part of the committee that compiled the new report. (14).

COP24 U.N. Climate Change Conference, December 2018 in Katowice, Poland.

COP24 is adopting the final agreement during the closing session of the COP24 U.N. Climate Change Conference 2018 in Katowice, Poland, on December 15, 2018. Nearly 200 countries overcame political divisions to agree on rules for implementing a landmark global climate deal, *"but critics say it is not ambitious enough to prevent the dangerous effects of global warming"*. A U.N. commissioned a report by the IPCC in October 2018, warned that keeping the Earth's temperature rise to 1.5 C would need *"unprecedented changes in every aspect of society"*. Saudi Arabia, the United States, Russia, and Kuwait refused to use the word *"welcome"* in association with the findings of the report. In other words, *"the Katowice Conference was again a failure"*. (UN Report).

COP25 United Nations Madrid Summit, 2-15 Dec. 2019 Climate envoys from almost 200 nations will gather in Madrid starting Monday, Dec. 2nd for two weeks (+ 2 extra days) of talks organized by the United Nations, to take new actions on climate change. A handful of major states resisted pressure to ramp up efforts to combat global warming. The COP25 talks in Madrid were viewed as a test of governments' collective will to heed the advice of science to *"cut greenhouse gas emissions more rapidly"*. But the conference, in its concluding draft, endorsed only a declaration on the *"urgent need"* to close the gap between existing emissions pledges and the temperature goals. Brazil, China, Australia, Saudi Arabia, and the United States had led resistance to bolder action, delegates said. *"These talks reflect how disconnected country leaders are from the urgency of science and the demands of their citizens in the streets,"* said Helen Mountford, Vice President for Climate and Economics, at the World Resources Institute think-tank. *"They need to wake up in 2021."* The country had earlier triggered outrage after drafting a version of the text that campaigners complained was so weak it betrayed the spirit of the Paris Agreement. In other words, the Madrid Summit was a failure too. The next Summit will be held in

Glasgow, the UK from Non. 1 to 12, 2021. Facts and reality presented in the next chapter prove *"the **INFERNO** in which Planet is entering"*. If the last three decades have taught the international community anything, it's that *"the science"* is not a single, settled entity that, presented properly, will spur everyone to action. There are no shortcuts to the technological, economic, political, and cultural changes needed to tackle climate change. That was true 30 years ago in Sundsvall. *"The only thing that has changed is the time in which we have left to do anything"*. And Mather Nation had and has shown the extreme weather events with much higher intensity and frequency, presented in the next chapter.

2

Extreme Weather Events

2.1. CLIMATE SCIENCE - GENERAL NOTES.

2.1.1. TIPPING POINTS

"A tipping point in the climate system is a threshold that, when exceeded, can lead to large changes in the state of the system." Potential tipping points have been identified in the physical climate system, in impacted ecosystems. Climate tipping points are of large interest in global warming nowadays. The tipping point has been identified for global warming. If several tipping points are reached, that will lead the world climate into a hothouse.

Climatically, the planet is divided into large areas, each area is characterized by sets of tipping points that create tipping elements. Tipping elements are found in the poles' ice sheets. The development of tipping points in time can be slow and persist for centuries to achieve endpoints. Some tipping elements, like the collapse of ecosystems, are irreversible.

For *"Climate Change"*, the *IPCC* definition for tipping points is *"an irreversible change in the climate system."* The tipping points can be determined by the rise of the temperature. For climate change, an *"adaptation tipping point"* has been defined as *"the threshold value or specific boundary condition where ecological, technical, economic, spatial or socially acceptable limits are exceeded."* A rapid or slow change in global average temperature can trigger *"global tipping points".* For instance, large-scale tipping elements are a shift in El Niño–Southern Oscillation. Reaching tipping points, the warm phase (El Niño) will continue in the southern ocean, which absorbs GHG, and can get in a state of no absorption. Also, climate change may trigger *"regional tipping points."* One best example is deforestation which triggers a tipping point in rainforests (i.e. Savannization in the Amazon rainforest, ...). There are several regional tipping points on the planet. The most critical is the burning of permafrost, due to the huge amount of carbon released into the atmosphere. Based on recent findings by researchers, tipping points are reached on rice at an average temperature of 1 to 2 degrees celsius.

Sometimes, reaching a tipping point may cause other tipping points to be reached. These are so-called *"cascading tipping points"*. Ice loss in the north and south poles will affect ocean circulation. Warming of the northern high latitudes will activate tipping elements in that region, such as permafrost degradation, loss of Arctic sea ice. This is an example of *"low levels of global warming, relatively stable tipping elements may be activated."*

A 2019 UNEP study indicates that now *at least for a study done by UNEP in 2019, "shows that the Arctic and the Greenland ice sheet tipping point has already been reached.*

The dewing of permafrost soil, releasing more methane in the atmosphere, and melting ice loss the reflecting power of solar rays lead to an increase of the temperature. This will accelerate climate instability in the polar region and affect the global climate. This phenomenon leads to *a "mass recession of Arctic sea ice"*. The example shows how year after year more ice melts. That has severe consequences for the entire planet. In June 2019, some of the fire reached peat soil, based on satellite data.

Peat is an accumulation of partially decayed vegetation and is an efficient carbon sink. The long-lasting peat fires release their stored carbon back to the atmosphere."*If the climate reaches into a greenhouse Earth scenario, warning of food and water scarcity, hundreds of millions of people displaced by rising sea levels, and so on"*.

There are a few scenarios, based on the level of temperature increase:

1). 4–5°C can make swathes of the planet around the equator uninhabitable, with sea levels up to 60 meters (197 ft) higher than they are today. Humans cannot survive if the air is too moist and hot, which would happen for the majority of human populations if global temperatures rise 2).11–12°C, the majority of humans can not survive. These effects are popularized in books like "*The Uninhabitable Earth and The End of Nature.*" Here is a list of tipping points analyzed by scientists (14).

1. Amazon rainforest, 2. Arctic sea ice, 3. Atlantic Circulation, 4. Boreal forests, 5. Coral reefs, 6. Greenland ice sheets, 7. Permafrost, 8. West Antarctic ice sheets, 9. Part of East Antarctica. (Wikipedia).

"*Warming process "triggering a cascade of tipping points or even to a global tipping point, and a less habitable, hothouse Earth.*" They are called "*world's climate tipping points, and we're getting close to the point of no return*". How might such a climatic collapse happen? Loss of Poles ice will drive freshwater into the North Atlantic. The phenomenon contributed to a 15% slowdown since the 1950s of the Atlantic Meridional Overturning Circulation (AMOC), a system that moves warm water northwards. "*Slowdown of the AMOC could destabilize the West African monsoon, causing drought in Africa's Sahel region. Also, a slowdown in the AMOC could dry the Amazon, disrupt the East Asian monsoon, heating the Southern Ocean, causing Antarctic ice loss.*" The West Antarctic ice sheet melting has passed a tipping point. "*The East Antarctic ice sheet might be similarly unstable, which could add another 3-4 meters to sea level on top of 3-4 meters from West Antarctic, a century or more from now. "Damaging tipping cascades may occur more than a global tipping point occurs, then this is an existential threat to civilization. Humanity has to change the approach to climate change*". A global rise in global surface temperature between 1°C and 2°C can tip the global tipping points. The world has already warmed by more than 1.2°C. Any action we might take to reduce global warming we will encounter at least 3°C. Speaking ahead of the UN's Climate Change Conference COP25 in Madrid, UN Secretary-General Antonio Guterres said: efforts to reach global targets have been "*utterly inadequate, and climate change could pass the point of no return.*" He emphasized that the world still has the means to keep warming below 1.5°C, but "*what is lacking is political will.*" (13). This statement is a diplomatic one to be optimistic and encourage the public. The way climate events evolve, and facts and reality tell us that the world can not stop warming below 1.5C, so *"we are doomed."* My opinion is backed by a group of influential scientists. (Scientists from Institute for Governance and Sustainable Development in Washington DC, and University of Exeter, UK). Changes in the frozen Greenland and Antarctica are *"dangerously close"* to tipping points. Fires in the Arctic and thawing ground permafrost will trigger a global tipping point where, *"whatever humanity does, there is no stopping it"*. The group of scientists says *"such a doomsday scenario is possible but more research is needed"*. Measuring tipping points and projecting when they will be reached is difficult, but others agree we are close to them. *"We may be at the precipice now concerning several key tipping points,"* says Michael Mann at Penn State University. *"Feedbacks and tipping points are the wild cards of the climate system, and they could push us into inevitable climate chaos far faster than heads of government understand,"* says Durwood Zaelke at the Institute for Governance and Sustainable Development in Washington DC. *"Climate Crisis, if we don't solve it soon, we will never solve it,"* because we will pass a series of irrevocable tipping points, and we're now approaching those deadlines. *If we wait past 2020 to reach the peak of GHG emission it's not a slope at all, it's just a cliff, and we fall off it"*. As the former UN climate chief Christiana Figueres put it when she launched Mission 2020, *"Everyone has a right to prosper, and if emissions do not begin their rapid decline by 2020, the world's most vulnerable people will suffer even more from the devastating impacts of climate change."* With other words: the IPCC reported last autumn, 2019, *"that if we hadn't managed a fundamental transformation of the planet's energy systems by 2030, our chance of meeting the Paris temperature targets is slim to none"*. The world's greenhouse gas emissions spiked last year, 2019, and given to Trump, Bolsonaro, Putin, Xi, and others, it's hard to imagine we have seen the same depressing thing this year, 2020.

Irreversible Change.

Johan Rockstrom is co-author of the *"Planetary Boundaries"* (2009) concept, a central paradigm for evaluating Earth's capacity to tolerate the impact of human activity, Rockstrom spoke with AFP, and here is an extract of his interview. Regarding a viable solution to global warming compatible with consumer capitalism. We don't have a choice. In the climate, time is running out so fast that there's no other pathway. Either we make this work within the existing economic paradigm, or we fail. It's naive to say *"Let's go for de-growth, let's completely divest, or let's think of a post-capitalist model and throw GDP in the waste bin".* We have to work with the economic machinery that we have in our engine room. The fossil fuel companies are reaching out to you, an Earth system scientist, for advice. Why? The European-based oil companies, such as Shell, BP, and Equinor, have not come to the point where they see the beginning-of-the-end of their business. Their argument goes like this: *"We serve humanity by supplying modern energy. Economies need cheap energy because otherwise they just grind to a halt. Don't blame us, we're just providing exactly what you are asking for."* As I will show in Chapter 3 companies are still making major natural gas and oil investments in the Arctic and elsewhere. Could Earth flip from being self-cooling to a self-warming system? The fact that the Earth system remains in net cooling mode. We are loading the planet with heat by burning fossil fuels. So far the Earth system has responded by dampening that heat.

*"**First,** land and oceans absorb 50% of emitted greenhouse gases."* That's the biggest economic subsidy ever.

*"**Second,** 90% of the excess heat we generate is taken up in the oceans".*

*"**Third**, ice sheets and the polar ice cap reflect 90% of solar radiation into space".*

We have carbon sequestration, heat absorption, and albedo, whereby the Sun's energy is reflected into space by mirror-like snow and ice, as the three big cooling agents on Earth. This is what has kept the system so resilient during the Holocene interglacial period (over the last 12,000 years). I call the planet's capacity to buffer abuse *"Earth resilience. The big nightmare is the moment when Earth can no longer cope and shifts from that self-cooling state to become a net-warmer."* The events and reality show the planet is in a warmer state. The Earth became a net source of greenhouse gases. *The point when the Earth system flips from being a net cooler to a net warmer is when you cross tipping points.* You could end up in a state where that resilience is weakened so much that the climate system no longer reduces human disturbance, but rather reinforces it. *"That will be the moment when we have lost control."* We have to start educating humanity that there are two-time scales that matter when it comes to climate change. **Educating humanity** is the basic subject of this book. Every conscious person on the Earth has the duty to educate herself/himself about the emergency crisis humanity has never encountered, but is experiencing nowadays. One is the deployment time-scale when we push the 'on' button of irreversible change, whether for the *"melting of ice sheets and the permafrost, or conversion of the Amazon from tropical forest to savannah"*, (due to deforestation). The other is the full impact time frame, which unfolds over centuries. *"We need to avoid the first one."* We don't want to push the buttons of runaway global warming." The next decade (2021-2030) is our window to avoid coming too close to those pressure points (15).

2.1.2. THE AGE OF STABILITY IS OVER

Humanity has only recently become accustomed to a stable climate. For most of its history, long ice ages punctuated with hot spells alternated with short warm periods. Transitions from cold to warm climates were especially chaotic. *"Then, about 10,000 years ago, the Earth suddenly entered into a period of climate stability modern humans had never seen before"*.

But thanks to ever-accelerating *emissions of carbon dioxide and other greenhouse gases, humanity is now bringing this period to an end. This loss of stability could be disastrous"*.

If the coronavirus pandemic can teach us anything about the climate crisis it is this:

"our modern interconnected global economy is much more vulnerable than we thought, and we must urgently become more resilient and better prepared for the unknown from climate change that can trigger more pandemics".

After all, a stable climate underpins much of modern civilization and its security. Extreme heat increases occurred in 2018. A prolonged heatwave and drought-hit much of western and northern Europe and decimated much of the potato harvest in the region. Temperatures in Germany reached record highs in a summer that was drier and hotter than in many parts of the Mediterranean. Only one year on, in 2019, western

Europe was struck by another *"impossible"* heatwave. In Germany, with temperatures topping 40°C, the record of the previous year was broken twice. Even in the Netherlands, known for its cool sea breeze at peak summer, high temperatures exceeded a searing 39°C.

Wildfires. A large part of Australia's forests is concentrated in the southeast of the country. In natural fires, typically 1-2% of the area is consumed by flames. Massive bushfires arrived much earlier than models predicted. Wildfire and climate models did predict a large increase in bushfire activity towards the end of this century. The models did not foresee that megafires will strike as early as 2019, and consume 20% of forests.

Locusts are a climate crisis. In the long term, the IPCC predicts *"crop yields will decrease by around 10%, but to date, it has ignored the possibility of large-scale pest outbreaks, which can wipe out entire harvests. In Kenya, it became the worst such outbreak for more than 70 years."* It is now feared that continued breeding of the locusts will create a second wave that will be far worse than the first one (16). Cornell Alliance for Science, June 4, 2020, By: Wolfgang Knorr).

Stopping Climate Change?

We have to think *"that it was 69F (20C) in Antarctica at the beginning of March 2020"*. And that bushfires consumed an area equivalent to the size of the United Kingdom in Australia, by one count, 1 billion animals were lost and 2 billion relocated. *"This is not about saving the planet"*. For one thing, the planet itself is not at risk, in its 4.5-billion-year history. *"It's civilization as we know it today that's in trouble"*. Second, *"the whole notion of "saving" anything is a flawed way to think about the crisis we are facing"*. We must eliminate fossil fuels and the faster we get off fossil fuels. *"But no matter how fast we act, we are not going to **fix** the climate"* like a doctor fixing a broken leg. ***"The Earth's climate is not a binary system or a switch that you can toggle on and off,"*** says Kate Marvel, a climate scientist at NASA's Goddard Institute for Space Studies in New York. *"Even if we stopped burning fossil fuels tomorrow and stabilized the Earth's temperature* *where it is today, we would still face several feet of sea-level rise"* in the coming centuries. We have already crossed one of the most important thresholds of the climate crisis: We've gone from *"Is it happening?"* to *"What are we going to do about it?"* In this new world. For this, there is no solution, only better and worse choices about where we will live, how we will live, who and what will survive. Above all, it's a world that will be defined by how hard we are willing to fight for our future. *"We might be living in a horror movie right now, but we are the ones writing the script,"* says writer Mary Annaïse Heglar. *"And we're the ones who will decide how this movie will end"* (NASA).

From the earliest days of the climate crisis, scientists have struggled to define the risks of life on a warming planet. *"We have understood the basic physics of climate change for more than 120 years,"* says Andrew Dessler, a climate scientist at Texas A&M. University. Unlike other air pollutants, such as the chemicals that cause smog, a good fraction of CO2 that was emitted while factories forged cannons during the Civil War (in America) is still in the atmosphere today, and will remain for centuries into the future. *"The climatic impacts of releasing CO2 will last longer than Stonehenge,"* wrote climate scientist David Archer. *"Longer than time capsules, longer than nuclear waste, far longer than the age of human civilization so far."* In *March 1958* the CO2 level in the atmosphere was 315.71 parts per million, (ppm). A year later, it was 316.71 ppm. In the week of 7-15th of November 2020, in New Delhi India the pollution gases reached 425 ppm. Big Oil and Big Coal understood the implications of rising CO2 levels all too well. They immediately began cranking out propaganda arguing that a warmer world was a better world. Groups like the Greening Earth Society argued that more CO2 meant plants would grow faster, agriculture would boom, and we would all enjoy more days at the beach. Companies like *"Exxon (now ExxonMobil) began spending hundreds of millions of dollars in a well-orchestrated campaign to deny, confuse, and block any understanding of the risks of burning fossil fuels."* As Dan Dudek, a vice president at the Environmental Defense Fund, puts it, *"What president or prime minister is going to restrict fossil fuels if it means he or she will be turned out of office?"*

In 1988, under the auspices of the U.N., the Intergovernmental Panel on Climate Change (IPCC) was created, an organization of top scientists tasked with issuing periodic reports that assessed the latest knowledge about climate change. The first report, released in 1990, was a weak sketch of the risks of sea-level rise, drought storms. Since then other summits have occurred, with not much more than words. In 1997, at the climate talks in Kyoto, Japan, UNFCCC members agreed to the *"Kyoto Protocol"*, which required that by 2012 developing countries cut total emissions of GHG by 5% from 1990 levels. *"Part of the problem was that negotiations focused on agreeing on the percentage of tons of carbon-dioxide-emission reductions, which no regular human being has any clue about,"* Dudek explains. *"How can you build political support around a goal that most people can't understand, even if they wanted to?"* (Need mass education).

Intensifying climate change pushed scientists to think about climate risk in terms of temperature changes, not carbon-emission rates".

In 2001, the IPCC issued its third report, which was far more pointed to a single graphic, known as *"the burning embers"* diagram. *"The diagram was revolutionary,"* says Michael Mann. *"For the first time, the risks of climate change were intelligible to someone who didn't have a degree in physics."* In 2010, the UNFCCC adopted a goal of stabilizing warming at less than 2 C (3.6 F), which quickly became known as the threshold for dangerous climate change. *Many pro and con discussions were about the 2C.* The counterargument, however, was that a climate target needed to be achievable or nobody would take it seriously. *"Virtually every study showed that hitting the 2C target would require a Herculean effort by all the industrialized nations of the world. Greenland and Antarctica are shrinking 100 years ahead of schedule,"* **There may be** a climate scientist or energy analyst somewhere in the world who believes that limiting warming to 1.5C is doable. *"Net emissions would need to fall by half by 2030".* How? When China says emission will pick up in China by 2030, the largest emitter on the planet. The emission has to be net-zero by 2050. *These targets are for dreamers.* *"The only analogy that has ever come close to capturing what's necessary is wartime mobilization, but it requires imagining the kind of mobilization that the U.S. achieved for less than a decade during WWII happening in every large economy at once, and sustaining itself for the remainder of the century."* In 2018, the IPCC published a special report that laid out the differences between a 2C world and a 1.5C world. *"I was grumpy about the idea of the 1.5C reports,"* says NASA's Kate Marvel. *"I* *thought it was just fanning fiction. But it had an unexpectedly galvanizing impact on people."* The report showed that, at 2C, severe heat events would become 2.6 times worse. 99% of coral reefs are dying, instead of 99% will die (17).

The latest climate models, which use more-sophisticated cloud-modeling techniques, *"are showing a higher climate sensitivity, with potential warming of as much as 5C if we double the CO2 in the atmosphere."* (At the current rate of emission of carbon, we double the quantity of GHG in 20 years, but the rates of emission will increase at least in the current decade 2021-2030. My calculation shows the doubling the amount of GHG will be reached in less than 15 years). *"The other big uncertainty about our climate's future has to do with tipping points."*

For example the **Gulf Stream** system, for example, *"has been slowing down in recent decades,"* says Gavin Schmidt, head of NASA's Goddard Institute for Space Studies. Melting of the **permafrost** in the Arctic: The more the permafrost warms, the more methane it releases, the more it warms the atmosphere. Similarly with the **Amazon rainforest**: As warming combines with deforestation, parts of it may turn into more of a *"savannah-like"* ecosystem. *"But it's not like there is a sudden crash and the entire Amazon disappears,"* says Hausfather.

Concerning the **West Antarctic Ice Sheet, it looks** more unstable. Earlier this year 2020, researchers in Antarctica found evidence of *"warm water directly beneath the glacier,"* which is not good news for the stability of the system. Eric Rignot, a scientist at NASA's Jet Propulsion Laboratory and one of the top ice scientists in the world, believes that the West Antarctic Ice Sheet is beyond its tipping point and in the midst of an irreversible collapse. As Rignot put it recently, ***"The fuse has been blown."*** Also, when you look at the **bushfires in Australia** (and all above), it's easy to think that *"it's too late to do anything about the climate crisis"*, that we are, for all intents and purposes, **in a catastrophic danger zone**, (not to use American expression f----d). *"But the lesson of this is not that we're doomed, but that we have to fight harder for what is left"*. Too Late-ism only plays into the hands of Big Fossil Fuels, and all the activists who want to drag out the transition to clean energy as long as possible.

Too Late-ism also misses the big important truth that, buried deep in the politics and emotion of the climate crisis, you can see the birth of something new emerging. *"The climate crisis isn't an 'event' or an 'issue,'"* says futurist Alex Steffen, author of *Snap Forward,* an upcoming book about climate strategy for the real world. *"It's an era, and it's just beginning"* (An era of centuries' time frame). According to a new poll from the Yale Program on Climate Change Communication, *"nearly six in 10 Americans are now alarmed or concerned about global warming."*

Germany, the industrial powerhouse of Europe, plans to shut down all coal plants by 2038. Sorry, but too late to help society attenuate the Climate Crisis. In the U.S the coal industry is in free-fall. Jim Cramer, CNBC's notoriously cranky Wall Street guru, said in January 2020, *"I'm done with fossil fuels. ... We're in the death-knell phase. ....The world has turned on them. It's kind of happening very quickly. I don't have any doubt that we will take action on climate,"* says Steffen. *"But it won't be the old-fashioned version of social change. It won't be an orderly transition. It won't be the climate version of the civil-rights movement. It will be more like the Industrial Revolution, a huge social and cultural, and economic transition, which will play out over decades, and with no clear leadership and nobody in control."* In Steffen's view, *"climate doomers are as blind as climate deniers. The apocalyptic is in its very heart a refusal to see past the end of an old worldview, into the new possibilities of the actual world"* (18). A scientific study published Thursday, November 12, 2020, concludes. "The world is already past a point of no return for global warming," reported in British journal Scientific Reports. The only way to stop the warming, they say, is that *"enormous amounts of carbon dioxide have to be extracted from the atmosphere."* (Later in the book I will make more analysis about direct air capture DAC). The scientists modeled the effect of GHG emission from 1850 to 2500 and created projections of *"global temperature and sea-level rise. According to our models, humanity is beyond the point of no return when it comes to halting the melting of permafrost using greenhouse gas cuts as the single tool,"* lead author Jorgen Randers, a professor emeritus of climate strategy at the BI Norwegian Business School, told AFP. *"The study said that by the year 2500, the planet's temperatures will be about 5.4F (3C) warmer than they were in 1850. And sea levels will be roughly 8 feet higher (2.45 meters)".* The authors suggest that global temperatures could continue to increase after human-caused GHG emissions have been reduced. According to the study, to prevent the authors' projected temperature and sea-level rises, all human-caused GHG emissions would have had to be reduced to zero between 1960 and 1970. *"To prevent global temperature and sea level rises, at least 33 gigatons of carbon dioxide would need to be removed from the atmosphere each year from 2020 onward through carbon capture and storage methods"*, which is the total amount of carbon dioxide the global fossil fuel industry emitted in 2018 (19). Mark Maslin, a professor of climatology at University College London, pointed to shortcomings in the model, telling AFP that the study was a *"thought experiment. What the study does draw attention to is that reducing global carbon emissions to neutral zero by 2050 is just the start of our actions to deal with climate change,"* Maslin said. The study authors urge other scientists to follow up on their work: *"We encourage other model builders to explore our discovery in their (bigger) models, and report on their findings"* (20). The following extreme weather events shall underline the importance of action needed to be *taken,* to reduce to the maximum the effects of GW.

2.2. EXTREME WEATHER EVENTS.

Nature struck relentlessly in 2020 with record-breaking and deadly weather- and climate-related disasters. With the most named storms in the Atlantic, the largest-ever area of California burned by wildfires, killer floods in Asia and Africa, and a hot, melting Arctic, 2020 was more than a disastrous year with the pandemic. The United States didn't just set a record for the most disasters costing at least $1 billion (per occurrence). The nation obliterated the record, according to the (NOAA). *"The 2020 disaster cost reached $96 billion."* Only three states weren't part of a billion-dollar weather disaster (Alaska, Hawaii, and North Dakota). With 30 named storms, the Atlantic hurricane season surpassed the mark set in 2005. A dozen made landfall in the U.S, and Louisiana got hit five times. At one point, the American Red Cross had 60 New Orleans hotels filled with refugees.

-With a devastating 20-year megadrought and near-record heat, *"California had at least 6,528 square miles (16,907 square kilometers) burned by wildfire"*, doubling the previous record area burned. Five of the six largest wildfires in California history have been in 2020. More than 10,000 buildings were damaged or destroyed and at least 41 people were killed.

Between fires and hurricanes, the American Red Cross provided a record 1.3 million nights of shelter for disaster-struck Americans, *"four times the annual average for the previous decade." "Since April, we've seen a large disaster occur somewhere in the country every five days,"* said Trevor Riggen, the Red Cross vice president. *"It's been a non-stop pace, and not all those disasters make the news."*

-The derecho storm that ravaged had damage estimated to $10 billion. Other billion-dollar severe storms, often with tornadoes and hail, struck the U.S. in January, February, twice in March, three times in April, and another three times in May 2020. All these U.S. disasters have *"really added up to create a catastrophic year,"* said Adam Smith, an NOAA-applied climatologist. *"Climate change has its fingerprints on many of these different extremes and disasters."* Nature is sending us a message. *"We better hear it,"* United Nations Environment Programme Director Inger Andersen told The Associated Press *(16a, NOAA & NASA).*

-**Worldwide,** more than 220 climates, and weather-related disasters hurt more than 70 million people and caused more than $69 billion in damage (not including the U.S.). Over 7,500 people were killed, according to preliminary figures from the international disaster database kept at the Centre for Research on the Epidemiology of Disasters at the Catholic University of Louvain in Belgium. Of the disasters, *"85% to 90% are climate and weather-related."*

-Worldwide floods are the biggest problem. It's a huge mistake to underestimate floods." Floods killed more than 1,900 people in India in June and affected 17 million people. Other flooding and associated landslides in Nepal, Pakistan, Afghanistan, and again in India killed at least another 1,250 people. African floods killed nearly 600 people. Flooding along the Yangtze River and the Three Gorges Dam in China killed at least 279 people in the summer and caused economic losses of more than $15 billion, according to the World Meteorological Organization (WMO). Vietnam had a record 103 inches (261. centimeters) of rain in October, according to the WMO.

-The Arctic was 9 F(5 C) warmer than average and had an exceptionally bad wildfire season. Arctic sea ice shrank to the second-lowest level on record.

-The pace of disasters is noticeably increasing, said, disaster experts and climate scientists. The international database in Belgium calculated that *"from 1980 to 1999, the world had 4,212 disasters affecting 3.25 billion* *people and costing $1.63 trillion, adjusted for inflation. From 2000 to 2019 those figures jumped to 7,348 disasters, 4.03 billion people affected and $2.97 trillion in damage".* Climate change figures in the growth of disasters, especially wildfires worsened by drought and heat, said Pennsylvania State University climate scientist Michael Mann. *"I didn't expect to see a season with 30 named storms in my lifetime,"* Mann said.

-Near-record drought and heat caused heavy crop losses in South America. Much of central Europe had an extensive drought, with a *"record 43-day spring dry spell in Geneva",* home of the WMO. *"These are the types of events scientists fear will increase in frequency, and intensity becomING the norm on the weather." (WMO, NOAA)*

2.2.1. DROUGHTS

Presenting a few examples of Megadrought and droughts around the planet, in the last year 2020, will emphasize the role of global warming effects on the destruction of droughts.

Megadrought' Emerging in the Western US. A few recent examples of droughts and Megadroughts on the planet. A *"megadrought"* appears to be emerging in the western U.S., a study published Thursday, April 16th, 2020, suggests. The nearly-20-year drought is almost as worse than any in the past 1,200 years, scientists say.

Megadroughts, defined as intense droughts that last for decades or longer, once plagued the Desert Southwest. *"We now have enough observations of current drought and tree-ring records of past drought to say that we're on the same trajectory as the worst prehistoric droughts,"* said study lead author, Mark Williams, a climatologist at Columbia University, in a statement. This is *"a drought bigger than what modern society has seen."* Scientists say that about half of this historic drought can be blamed on man-made global warming. Some of the impacts today include shrinking reservoirs and worsening wildfire seasons. The drought will continue for the near future, or fade briefly, researchers say. The study covers an area stretching across nine U.S. states from Oregon and Montana down through California, New Mexico, and part of northern Mexico (20).

Colorado Drought. Every part of the Centennial State is presently under drought or abnormally dry conditions for the first time in nearly a decade, according to the National Drought Mitigation Center. Experts say the drought conditions. Global warming *"continues to stress the region as a whole."* In Colorado, the consequences are most keenly felt in the agricultural and water management sectors, according to Peter Goble, a climatologist at the Colorado Climate Center at Colorado State University. Currently, 26.6% of Colorado is characterized as being under "extreme drought." This classification is characterized by the potential for major crop and pasture losses and widespread water restrictions. More than 32% of the state is under "severe drought" conditions, 27% under "moderate drought" conditions, and 14.% characterized as being "abnormally dry," according to the NDMC's drought monitor map, which was last updated on Aug. 6, 2020. All these droughts led to extreme fire as presented in the next paragraph (21).

Chile Drought. Chile saw an 80% drop in rainfall in 2019 compared to its previous historic low.

With historically low river flows and reservoirs running dry due to drought, *"people in central Chile have found themselves particularly vulnerable to the coronavirus pandemic".*

Years of resource exploitation and lax legislation have allowed most reservoirs in that part of the country to run dry. *"There are now 400,000 families, nearly 1.5 million people approximately, whose supply of 50 liters of water a day depends on tankers."* One of the main pieces of advice to protect people against coronavirus is to wash your hands regularly. *"Living without water is awful".* In the greater Santiago area and in Valparaiso, rainfall last year (2019) was almost 80% below the previous record low. In the northern region of Coquimbo, it was down 90%.

Water tankers serve many homes, whose inhabitants come out to fill drums.

The virus pandemic is highlighting *"once more that where there is a model of the private appropriation of water ... this condition does not guarantee people's human right to water and further weakens communities."* The Melon river in Chile was completely dried out in January 2020. Chilean law states that water is a resource for public use, but it turned over almost the entirety of the right to exploit the resource to the private sector (22).

Drought Parches Southern Africa, Millions Faced With Hunger

An estimated 45 million people are threatened with hunger by a severe drought strangling wide stretches of southern Africa. Food deliveries are planned for Zambia, Zimbabwe, and other countries hard-hit by a combination of low rainfall and high temperatures. *"We are witnessing millions of already poor people facing extreme food insecurity.* Parts of Zimbabwe have had the lowest rainfall since 1981, contributing to putting more than *"5.5 million at risk of extreme food insecurity",* Oxfam said in a report. In Zambia 2.3 million people are food-insecure.

The drought is also worsening food availability in Angola, Malawi, Mozambique, Madagascar, and Namibia. Southern Africa has *"received normal rainfall in just one of the past five growing seasons".* Two cataclysmic cyclones hit Mozambique, Zimbabwe, and other southern African countries early this year (2019). *"The successive mixture of drought and flooding has been catastrophic for many communities. In most of the affected areas, there isn't enough drinking water, which means that people and animals, both livestock and wildlife, are having to use the same water points. This is unacceptable as it exposes people to diseases and creates a heightened risk of wild animal attacks".* The drought has also affected the region's wildlife. At least 105 elephants have died in Zimbabwe as a result of lack of water and vegetation, according to Zimbabwe's National Parks (23).

2.2.2. FIRE AND FIRENADO

The weeks of August 8th to 23rd, have been hell in California and the West Coast of America. With Megafire, Fire Tornado, the most polluted San Francisco & Bay Area in the world, everything has been Apocalyptic. The West Coast has experienced the most devastating fire in the history of the United States, if not of the Modern World. "*The fire experienced in California will become the norm for years ahead.*"

A Blink Looks to The Last Year (Sept 2019 to Oct. 2020) Fires.

A good, natural fire can be a cleansing force. Yet, the recent and ongoing catastrophic fires around the world – including in Brazil, the US, Sweden, Russia, and Australia, are not moments of a healthy fire cycle but conflagrations of a dying biosphere. Terrible as they are, the fires in the western American states are only middling on a global scale. As of early November, 8.6millions acres (3.5mil. hectares) had burned nationally, with half of that total in California. But the American catastrophe pales in comparison to Australia's wildfires in 2019, which incinerated an eye-watering 46 million acres. More than 20% of the country's forests were destroyed in a single year. Siberia's fires in 2020 were even bigger, 47 million acres. A tenth of South America's largest wetland, the Pantanal, went up in smoke this year 2020, some 6 million acres, coupled with the Amazon losing 8.5 million acres. That latter figure is only half the size of last year's (2019) fiery nightmare.

Paradise Fire "Campfire"

"Everybody at some point was certain they were going to die: Paradise's former mayor recalls devastating wildfire two years on (27 November 2018)".

Steve *"Woody"* Culleton was driving to work 20 miles away from his home in Paradise, California when he spotted a cloud of smoke. *"I happened to have a police scanner, so I turned it on. By 7.12 am, the fire was in Paradise behind the hospital. The winds were blowing 40-50 miles an hour,"* he told The Independent. *A home burns as the CampFire rages through Paradise, California on Thursday, November 8, 2018 (Noah Berger).*

Mr. Culleton, now 76, raced home to find that *"fire was already in the neighbor's yard"*. Within minutes thousands of residents in Paradise were trying to escape an out-of-control wildfire. *"I met my wife at the house and tried to grab our pets but we weren't able to corral them,"* Mr. Culleton recalled. *"We grabbed a few things and got in separate vehicles. I went to pick up our adult daughter and my wife continued, trying to get out of town."*

The family lost their home and possessions in the devastating firestorm, named CampFire. By the time it was extinguished, 85 people had lost their lives, the deadliest blaze in a century in the US. Some 50,000 people were displaced and 95% of the structures in the 141-year-old town were destroyed. *"With almost 6,000 first responders battling the fire, it was finally contained 17 days later. The damages ran to $16bn".*

The story of how the fire impacted the small community, and the arduous recovery that followed, is the subject of Rebuilding Paradise, a National Geographic documentary from Ron Howard, the Oscar-winning director of movies A Beautiful Mind and Apollo 13.

It premiered in the UK earlier this week (23 to 29, Nov 2020) and in the US on the second anniversary of the tragedy". The documentary begins with heart-stopping footage contributed by Paradise residents fleeing the fire. Howard's film crew spent a year in Paradise following the grieving community as residents faced relocations, financial stress, bureaucratic red tape, water poisoning, and post-traumatic stress disorder. *"When we went to Paradise, the range of emotions was so palpable on people's faces and in the words of the people I was meeting, some of whom we were rolling cameras on, and some of whom we were just speaking to grasp it all,"* the director said. *"I have never been around anything like that kind of devastation. It was jaw-dropping and heartbreaking."* Mr. Culleton was one of several local people to appear in the film, along with Officer Gates, who told of watching his home burn as he evacuated people. There's also school superintendent Michelle John, whose husband Phil died while the film was being made, and Carly Ingersoll, a school psychologist who helped students cope with the trauma. Mr. Culleton said that he felt the government let people down in the immediate aftermath. *"The government wouldn't let people in, wouldn't tell anyone anything. We just did things that the government failed to do."* Resources from FEMA are stretched thin, the film captures the community battling against state government bureaucracy, along with their flagging spirits. Before the wildfire, around 26,000 people lived in Paradise. Two years on, Mr. Culleton estimates that some 4,000 people have come back. But for many, it has been impossible to return to Paradise. *"Everybody that experienced that day, at some point in their evacuation was certain they were going to die,"* Mr. Culleton said. *"I know friends that have not been back to town since the day of the fire. It's too traumatizing, every time the wind blows.... Many of us don't sleep hardly at all anymore, maybe four hours a night. People have moved away – Idaho, Texas, the East Coast, Oregon."* The wildfire was ignited by Pacific Gas and Electric Corp's faulty electrical grid which the company had failed to maintain. In June 2020, PG&E, the US's largest utility provider, pleaded guilty to 84 counts of manslaughter for its role in CampFire. The company declared bankruptcy during which a $25.5bn settlement was worked out. In July 2018, the fire in Redding killed eight people including three firefighters, and nearly exhausted the resources of the statewide fire department Cal Fire. The area had experienced five years of drought. Xan Parker, the producer of Rebuilding Paradise, explained: *"Our co-producer Lizz Morhaim said, 'PG&E lit the spark, but climate change laid out the tinder. And the tinder is everywhere, waiting for a spark."* Southern California has *"warmed about 3F (1.67C) in the last century."* 2020 has seen unprecedented wildfires in California, that have obscured the sun and turned skies an apocalyptic orange. On the morning of the fire, and evacuating resident manages to capture footage of the town sign, as it catches fire. It reads: *"May You Find Paradise To Be All Its Name Implies. This fire has changed everybody's life. We are all different today as a result of that fire"* (25). The movie is a must-see.

Australia Fire 2019. Australia's fire is in the grip of deadly wildfires burning across the country, triggering an emotive debate about the impact of climate change in the world's driest inhabited continent. The unprecedented scale of the crisis, and images of terrified tourists sheltering on beaches from the infernos, have shocked many Australians, and the world. With the nation only midway through summer and suffering a prolonged drought, authorities fear the death toll will continue to mount as more homes and land are destroyed. Here are some key details of the crisis:

How many people have died?

Since the fire season began months ago during the southern hemisphere winter, at least 33 people have died. Among the fatalities are volunteer firefighters, including a young man who died when his 10-ton truck was flipped over in what officials have described as a **"fire tornado."** similar to the one developed in California in August 2020. Australia's worst wildfires came in *"2009 when the Black Saturday blazes left 180 people dead"*.

How big an area has burned?

More than 30 million acres (12.5 million hectares) have been destroyed, which is more than 4 times the size of Wales. In New South Wales state alone, 8.9 million acres of forest and bush have been destroyed, while more than 2.2 million acres have been burned in Victoria. The fires are so large they are generating their weather systems and causing dry lightning strikes that in turn ignite more. One blaze northwest of Sydney, the Gospels Mountain fire, has destroyed more than 1.2 million acres, about seven times the size of Singapore.

What's the economic impact?

The Australian Open, the first Grand Slam of 2020 that brings in an estimated A$290 million, has seen qualification games disrupted due to the smoke. Economists estimate the wildfires and associated drought could cut up to half a percentage point off GDP growth as agriculture, tourism and sentiment take a hit.

How many homes have been destroyed?

Some 2,600 homes have been destroyed, mostly in New South Wales. Scores of rural towns have been impacted, including the community of Balmoral about 150 kilometers southwest of Sydney, which was largely destroyed before Christmas.

How has wildlife been affected?

The University of Sydney estimates that 800 million animals have been killed by the bushfires in New South Wales alone since September 2019, and one billion have died nationally. The *"highly conservative figure"* includes mammals, birds and reptiles killed either directly by the fires, or later due to loss of food and habitat. The fires have raised concerns in particular about koalas, with authorities saying as much as 30% of their habitat in some areas had been destroyed. Images of the marsupials drinking water from bottles after being rescued have gone viral on social media (26).

Climate change lengthens Australian summers by 50%.

Australian summers are now effectively twice as long as their winters as climate change has increased temperatures since the middle of the last century, research released in the wake of the nation's unprecedented fire season showed. The report by the Australia Institute, a Canberra-based think tank, compared data from the past two decades with mid-20th century benchmarks of temperatures. Over the last two decades (2000-2020), summer across most of Australia has been on average one month longer than half a century ago, one more strange climate phenomenon. *"Over the past five years, the analysis showed, Australian summers were on average 50% longer than they were in the mid-twentieth century,"* based on temperature readings. *"This is happening right now."*

Wildfire Emergency Declared as Heatwave Grips Australia

"The biggest concern over the next few days is the unpredictability," Sydney was again blanketed with a toxic haze as smoke billowed in from the fires. The severity and unusually early start to the fire season, has intensified the *"debate about what impact climate change is having on the world's driest inhabited continent"*. The worst of the heat is in the continent's interior, with vast swathes of territory experiencing

severe heatwave conditions. Adelaide is forecast to reach 45C (113F) on Dec. 18, 2019. while the nation's capital Canberra is bracing for three consecutive days above 40C. Since the fire season began, the dust and particles from the blazes have reached the glaciers of New Zealand, staining them red. Many social media users and some news outlets *"criticized Prime Minister Scott Morrison for taking an unannounced vacation with his family to Hawaii just days after labeling the blazes a national disaster."* **Morrison's (a denier of climate change)** conservative government has been a strong advocate for the coal industry and steadfastly *"opposes putting a price on carbon"* (27).

Another California Fire Tornado In New Sierra Blaze.

California officials issued a warning for a "fire tornado", or "firenado", in a new blaze in the eastern Sierras. *"The wildfire is burning so hot that it has the power to create its weather phenomenon with deadly blasts of blazing heat, flames, and wind."* The wildfire raging near the small community of Loyalton near Reno, Nevada, grew to 20,000 acres by Saturday, August 15th. *"Some 300 firefighters"* were battling the blaze, which was first reported Friday, August 14th, afternoon. It was only 5% contained the next day. *"Fire behavior is extreme, and a smoke column will be visible throughout the Sierra Valley and North Reno."* Officials spotted a threatening *"pyro-cumulonimbus cloud"* in the early afternoon created by the fire south of Chilcoot. The rotating columns and fire whirls of the burgeoning cloud are *"capable of producing a fire-induced tornado and outflow winds over 60 mph,"* the National Weather Service warned in its fire tornado alert. *"This is extremely dangerous for firefighters."* Extremely dangerous fire behavior noted on the LoyaltonFire! Rotating columns and potential for fire whirls. Responders should exercise extreme caution!!! The tornadic pyrocumulus has weakened, and the immediate threat of tornadic activity has decreased for the Loyalton Fire. Even so, please stay away from the fire area. There were no immediate reports of injuries, but sections of three counties were under mandatory evacuation orders (28).

California Wildfires Spawn First "GIGA FIRE" in Modern History

August complex fire expanded beyond 1 million acres, elevating it from a mere 'megafire' to a new classification: 'gigafire' California's extraordinary year of wildfires has spawned another new milestone, the first "gigafire", a blaze spanning 1 million acres, in modern history.

At 1.03 million acres, the fire is larger than the state of Rhode Island and is raging across seven counties, according to fire agency Cal Fire. An amalgamation of several fires was caused when lightning struck dry forests in August, 12,000 strikes in 72 hours, the vast conflagration "***has been burning for 50 days and is only half-contained***". The August complex fire heads a list of huge fires that have chewed through 4 million acres of California this year 2020, a figure called *"mind-boggling"* by Cal Fire and doubles the previous annual record. *"Five of the six largest fires ever recorded in the state have occurred in 2020,"* resulting in *"several dozen deaths and thousands of lost buildings"*. Vast, out-of-control fires are increasingly a feature in the US west due to climate change (California experienced in the last decade a Giga drought too). Big wildfires are three times more common across the west than in the 1970s, while the wildfire season is three months longer, according to an analysis by Climate Central (29).

"We predicted last year that we were living with the chance of such an extreme event under our current climate," said Jennifer Balch, a fire ecologist at the University of Colorado Boulder. *"Don't need a crystal ball."* The 2020 fire season has caused choking smoke to blanket the west coast and at times blot out the sun. But experts warn *"this year may soon seem mild by comparison as the world continues to heat up due to the release of greenhouse gases from human activity. If you don't like all of the climate disasters happening in 2020, I have some bad news for you about the rest of your life,"* said Andrew Dessler, a climate scientist at Texas A&M.

University. Much of the Central Valley is still under an air quality alert because of wildfire smoke from the Creek Fire, which has burned more than 326,000 acres, and the SQF Complex fire, which has burned nearly 159,000 acres in the Sierra National Forest. Northwest California, where the August Complex fire rages, had air quality "in the unhealthy to the locally hazardous category" as well (30).

About 560 Wildfires Burn in California

More than 100,000 people evacuated: Scorching heatwave burning for two weeks ignited hundreds of wildfires in California. At least six people have died. The state is battling about two dozen large complex fires. In total, 560 fires are raging in the state. There, fires have chewed through brushland, rural areas, canyon country, and dense forest surrounding San Francisco. Fires have charred over 1,200 square miles across the state, according to Cal Fire, a size comparable to Rhode Island. *"We're putting everything we have on this,"* Newsom said, noting the state is seeking aid from the federal government, other states as far away as the East Coast, and even Canada, Israel, and as far as Australia. Two of the fires there are now the 7th and 10th largest fires in state history, having burned as much as 300 square miles each.*"If you are in denial about climate change, come to California,"* More than 12,000 firefighters are on the front lines. Some firefighters were working *"72-hour shifts instead of the usual 24 hours."* The Northern California fire has now *"claimed at least six lives and injured 33 people and firefighters".* *"In central California, a pilot on a water-dropping mission in western Fresno County died."* Wednesday (August 5th, 2020). 119,000 people have already evacuated and 140,000 are under evacuation orders. The University of California, Santa Cruz was evacuated, and a new fire burning near Yosemite prompted residents to flee.

California is on fire: What are fire whirls, fire tornadoes, fire clouds, and running crown fires?

Of the 560 fires burning throughout the state, dozens of fires have collected into major complex fires. Many were sparked by *"an unprecedented dry lightning siege of nearly 12,000 strikes."*

Triple-digit heat and monsoonal moisture pulled from the south. Up to 20 separate fires are burning near San Jose, named the SCU Lightning Complex Fire. A new fire burning in California is the LNU Lightning Complex, which has burned a total of 342 square miles. Smoke from the fires makes pollution in California the worst in the world. *"Whenever things burn there's a mixture of gases and particles,"* said University of California, San Francisco pulmonologist Dr. John Balmes, an expert on the respiratory and cardiovascular effects of air pollutants. *"When you smell smoke, you are smelling the gases in the air, not PM 2.5,"* Balmes said. Limiting outdoor activities, remaining indoors with the windows and doors closed, with your AC open may help reduce your risk if you live in an affected area (31).

We Are Amid a Climate Emergency.

Even in a year of record wildfires, the *"August Complex Fire" north* of San Francisco is staggering, burning from the first week of August 2020, until Sept. 12ᵗʰ, 2020. This one is 60% larger than the 2018 MendocinoFire Complex that blackened 459,123. *"5 of the top 20 largest wildfires in California history have occurred in 2020"*. The August Complex Fire on August 12, had become the biggest in California's recorded history to over 746,000. acres burned. The August Complex merged with the nearby Elkhorn Fire is one more huge fire in CA. The North Complex Fire swept through the small town of Berry Creek, *"killing 10 people and leaving 16 missings."* Earlier on August 13ᵗʰ, California Governor Gavin Newsom toured the North Complex fire zone. Newsom called "BS" on climate deniers and said bluntly, "We're in the midst of a Climate Emergency, and California Declares Statewide Emergency in August 2020 (32).

The power cuts come as firefighters battle more than two dozen large wildfires across the state, Many fires sparked by dry lightning, strong winds, and triple-digit temperatures. In Death Valley, California, temperatures hit 130 F (51.4 C) degrees, *"which would be the hottest temperature recorded on Earth since at least 1913. The unrelenting heat has created a challenge for firefighters, in particular". "They're wearing a lot of gear, carrying equipment, hiking to remote locations, so it's a stress on the body,"* California's governor declared a statewide emergency on Tuesday, August 18ᵗʰ. This assures assistance from the Federal Emergency Management Agency (FEMA) to respond to fires in Napa, Nevada, and Monterey counties. Extreme events of climate are likely the *"new norm, this is exactly what so many scientists have predicted for decades"* (32).

California Wildfires, by the Numbers, as of Sept. 17, 2020.

After 45 days of burning California, there are still four months left of the fire season. Fires are raging across the entire West Coast, in California, Oregon, and Washington.

- 3.4 million acres have burned so far in 2020. Last year, 259,823 acres were burned.

- 25 people are confirmed to have died since fire activity elevated on 15 August.

- California currently has 25 major fire complexes burning around the state (Sept. 17th, 2020).

- There have been nearly 7,900 fire incidents since the beginning of 2020.

- Insured losses are almost $5 billion so far in 2020, according to Moody's Investors Service. -17,000 firefighters are tackling the blazes on the frontlines.

- More than 38,000 people have been evacuated around the state.

-Today 17,000 firefighters are battling 25 major wildfires statewide.

- Some 5,792 structures have been damaged or destroyed.

- Creek Fire in Fresno County has *"uniquely challenging conditions"*. The area has 163 million dead trees, acting as fuel, due to historic drought (Megadrought).

- 10 of the 20 most destructive California wildfires were in the last 5 years.

These data are for California. Similar situations encounter the West Coast states: Oregon, Washington, plus Nevada, Idaho, and Colorado. The San Francisco - Bay Area was the most polluted area in the entire world. The West Coast's wildfires show as another horrifying reminder of the effects of climate change on our planet. Those apocalyptic photos of San Francisco go across social media (33).

That was the Mendocino Complex Fire. While the area still has burn scars from the 2018 blaze, the August Complex Fire has raged for weeks across the forests of Mendocino County.

More Blaze In Sonoma And Napa

The LNU Complex fire in Northern California became the *"second-largest wildfire in state history overnight, according to Cal Fire."*(August 22, 2020). The blaze has scorched 314,000 acres in five counties: Sonoma, Lake, Napa, Yolo, and Solano. It is second in size to the Mendocino Fire in 2018. The CZU complex fires near Santa Cruz and the nearby SCU complex making it the 3rd largest fire in the state's history. The 4th largest wildfire in the state's history is the 2017 Thomas Fire in Ventura and Santa Barbara counties. The LNU complex fire has destroyed 560 structures, threatens another 30,000. and caused 3 deaths. Multiple fires have merged on the north side of Lake Berryessa into the Hennessey Fire. A *"complex fire"* is a network of fires burning in a given area that may or may not merge. *"The LNU Complex includes the Hennessey Fire, which is burning in Napa and Sonoma County. That blaze merged with the Gamble, Green, Aetna, Markley, Spanish Morgan, and Round fires. LNU also includes the Walbridge Fire in Sonoma County, which merged with the Stewarts Fire. "Big Basin is California's oldest state park and home to the largest continuous stand* of ancient coast redwoods south of San Francisco". In CA. "*We have a federal request in…a bipartisan request to get a major disaster declaration here in the state of California.*" The CA. state was battling 560 blazes and responded to 56,000 wildfires in 2020. My question to the reader is:" *How can I describe 56,000. of the fire which burned in a timeframe of 8 months*"?. The season fires last in 2020, until the last week of November (34).

Tiny California Town Leveled by "Massive Fire"

The tiny town of Berry Creek was demolished by the North Complex Fire near Oroville in Northern California. "*We had a massive wall of fire that came in and did everything it could to destroy everything in its path and unfortunately, there were some communities right in the way, just like what happened in Paradise and Concow during the Camp Fire,*" Cal Fire spokesman Rick Carhart said Thursday, Sept. 10, 2020. The Butte County Sheriff indicated that 10 people who died at 16 are still missing. Thus far the blaze has blackened 247,358 acres. It is being fought by over 1600 fire personnel (34). According to Cal Fire, "*an eight-mile-long flank of the blaze, previously known as the Bear Fire, swept across the hilly terrain and engulfed the towns of Berry Creek*". Firefighters engaged in lifesaving and evacuation operations, rescued over 100 residents, including burn people. "*I've only seen three homes left standing,*" Sacramento Bee photographer Jason Pierce told the paper. Every house is just dust. The same tragic events continue to sweep through wine country in Northern California, Nevada, and Colorado (35).

Even in December 2020, CA. was on fire. Out of the United States, I wish to present the fire with deep consequences on Climate Change, from Amazon forest, Brazil, permafrost in Siberia, and northern territories of Canada, and Greenland.

Amazon Fires: Year-on-Year Numbers Doubled in October 2020.

The Institute of Space Research said there were 17,326 fires in the Amazon in 2020, compared to 7,855 in October 2019. There were a record number of fires in the Pantanal wetlands. INPE data release shows there were 2,856 fires in the Pantanal region in October 2020, the largest monthly figure since records began over 30 years ago. President Jair Bolsonaro has not commented on the latest figures but has previously dismissed data from INPE as flawed. He also said the number of fires was *"disproportionate"*. (He likes his friend Trump, another denier of climate change). According to INPE, *93,485 fires* have been recorded in the Amazon so far this year 2020 - 25% more than in the same period in 2019, *"when President Bolsonaro's handling of the forest fires drew international condemnation."* A steep rise in fires in the Pantanal, one of the world's most biodiverse areas, home of jaguar and the capuchin monkey (36). The reader needs to know that the Amazon forest provides the planet with 20% of OXYGEN supply, and is considered as the plums of the Earth. The entire Amazon forest must become an *"International Planetary Environmental Habitat"*, and have similar policies and regulations as *"International Waters"*. This area is of life interest for the entire planet population, *"and has NOT to be exploited per the pleasure of DollarMan"*.

Brazil's Wetlands Are Engulfed in Flames.

The Pantanal, the world's biggest tropical wetlands, is burning at a record-shattering pace this year, 2020, in the drought-fueled fires. The region, located at the southern edge of the Amazon rainforest, is known for its immense biodiversity with *"its jaguars, jabiru storks, giant otters, caimans, toucans, macaws, and monkeys."* But in recent months, the images emerging from the region have been of *"charred animals' corpses and flames stretching clear across the horizon. I've been here 20 years, and this is the worst situation I've ever seen,"* Felipe Dias, head of the environmental group SOS Pantanal, told AFP. Stretching from Brazil into Paraguay and Bolivia, the Pantanal is crisscrossed by rivers, swamps, and marshes. More than *"2.3 million hectares (5.7 million acres), an area 388 times the size of Manhattan, have gone up in flames in the region so far this year 2020"* according to the Federal University of Rio de Janeiro. *"There have been 12,567 fires in the Brazilian Pantanal in 2020"*, setting a new annual record according to satellite data collected by Brazil's national space agency, INPE. *"Very few animals survive. The ones that do often suffer very severe effects. They're burned to the bone, they often have to be euthanized, or die of hunger and thirst"*. The worst part is when the people tell us, *'There's nothing we can do, everything is going to burn.' The only hope is for it to rain, but that's not expected until November."* Local volunteers have rushed to help the teams of soldiers and firefighters deployed to battle the flames. Many of them depend on the region's ecotourism industry. The world's biggest tropical wetlands are not that wet these days, due to climate change. Rainfall in the Pantanal plunged by 50% for the rainy season.

Studies show deforestation in the Amazon is having an impact on rainfall in other regions of Brazil by shrinking the rainforest's so-called *"flying rivers": vast clouds of mist that are carried by the wind and dump water across a large swathe of South America. But there's no denying things are different from before. I'm from this region. I remember when it used to rain in August and September. This year, it hasn't rained since June."* Droughts like this year's risk becoming the *"new normal,"* said Tasso Azevedo, the coordinator of Mapbiomas, a collaborative research group that tracks environmental data. *"That would be really tragic. Because in the Pantanal, if you have fire after fire in the same place, the vegetation can't grow back."* Destroying one of *"the most iconic ecological places in the world, is not only tragic, is a criminal act, irresponsible, irreparable, and desavouer."* Bolsonaro has to be named Climate Criminal and International Community to act on Amazon as a sacred area of the Planet. *(37).*

Our House is Burning

Emmanuel Macron, France's President, has dubbed the fires in the *"Amazon rainforest"* an *"international crisis"* and said it should be discussed at the upcoming G7 summit in 2019.

The French President, who is hosting the G7 summit, tweeted: *"Our house is burning. Literally."* The Amazon rainforest, the lungs of the plant, are on fire. *"It is an international crisis. Members of the G7 Summit, let's discuss this emergency first order in two days!* (August 2019). His comments come as data published by the National Institute for Space Research (INPE) revealed that fires across Brazil are up 85% this year 2019, many in the Amazon. (Note: In 2020 until Dec., the fire increased 12% more than in 2019). But Brazilian President Jair Bolsonaro accused Mr.

Macron of using the issue for *"personal political gain. He said: "The French president's suggestion that Amazonian issues be discussed at the G7 without the participation of the countries of the region evokes a misplaced colonialist mindset, which does not belong in the 21ˢᵗ century."* Amazon issues, as an *"International Crisis"* must be discussed at the United Nations Security Council, with the participation of all UN member countries.

Forest Fire Emissions From Indonesia Worse Than Amazon, EU Says.

South Asia fire, due to drought and set up by farmers, may have *"released the equivalent of 709 million tons of carbon dioxide through Nov. 15, 2020,"* about the same as the annual emissions of Canada, or 0.167% of global emission in 2019. *"That's 22% more than the estimated 579 million tons emission of Amazon forest fire"*, according to the Copernicus Atmosphere Monitoring Service, a program run on behalf of the European Commission.

The **Indonesian** fires, burned forest spread across *"more than 4,000 square kilometers, or about 2,500 square miles, with the worst-affected areas on the islands of Sumatra and Borneo"*. The haze from the fires affected air travel and caused respiratory illnesses for thousands of people. Was It Another Bad Year for Indonesia Forest Fires? Drought, Wildfires Inflict Double Whammy on Indonesian Crops. Emissions from Indonesia were the worst in the country since *"1,286 megatons of carbon dioxide equivalent were released by fires in 2015"* (3% of global emission in 2019), according to Copernicus. The Indonesian government estimated roughly 2.6 million hectares (6.4 million acres) of land was burned in 2015, costing the country $16 billion in economic losses. The blaze affected more than 850,000 hectares this year 2020, (2.04 million acres) an area equivalent to Yellowstone National Park in the U.S (37a).

SIBERIA, a Smoldered Zombie Fire Beneath Arctic Ice All Winter Then Reignited.

Fire causing one of the region's worst fire seasons in history
As temperatures rise in Arctic regions like Siberia, researchers have

observed *"zombie"* wildfires: blazes that smolder underground in the winter and return to the surface in the summer. If they combine with new blazes, making fires stronger and more widespread.

Fires in the Arctic region burn peat and other carbon-rich materials, releasing hundreds of millions of tons of carbon dioxide and methane into the atmosphere. In Eastern Siberia last year, 2019, summer fires burned more than *6.4 million acres in July 2019 alone* (2.67 million hectares), more than 2 million acres continued to burn through August. "*The blazes' dormant period has earned them a nickname: Zombie fires*". "*We have seen satellite observations of active fires that hint that 'zombie' fires might have reignited,*" Mark Parrington, a wildfire expert at (CAMS), said in a press release. "*Already, 2020 has brought one of Siberia's worst wildfire seasons on record*", according to Greenpeace. An area larger than Greece has burned (38).

The Arctic recent zombie fires are cause for concern because climate change is leading the Arctic region to warm at twice the average rate of the rest of the world. Siberia is at least 5.4F (3C) above its long-term average this year, 2020. In May, some regions saw temperatures 46F (7.7C) above normal, according to the Washington Post. A major heatwave in June drove temperatures in the Siberian town of Verkhoyansk to *"top 100 F (41C), possibly the highest recorded within the Arctic Circle"*. Zombie fires, in turn, add a *"cumulative effect"* to newer fires, Parrington said. "*The warmer it gets, the more fire we see.*"

In 2019, fires within the Arctic Circle released "*182 million metric tons*" of CO2, (A metric ton is about 2,200 US pounds.) Then in the first eight months of 2020, the fires released "*244 million metric tons*" of CO2, an increase of 34%, according to CAMS. *"In this part of Siberia, the signs of climate change are already here. It's not some distant future. It's now,"* Amber Soja, a research fellow at NASA and National Institute of Aerospace, told NASA Earth Observatory. *"But what happens in the Arctic doesn't stay in the Arctic, there are global connections to the changes taking place there. The world's permafrost contains over "1 trillion metric tons of carbon",* almost 4 times the existing GHG in the atmosphere, according to a recent analysis by researchers at the American Geophysical Union. *"Nowadays climate change models don't take this carbon into their projections.* This represents an underestimation of climate change. Deposit of darker soot which absorbs more heat leads to faster melting (39).

Siberia is Burning.

Scientists say Siberia is warming at twice the global average, leading to extreme weather events, severe environmental deterioration like permafrost burning, and serious complications for human habitation. And, they warn, *"it is a climate catastrophe that might be just beginning. If these temperatures repeat themselves next year, the situation on the southern fringe of Siberian forests is going to become critical,"* says Nadezhda Chebakova, a researcher at the Sukachev Institute of Forest in Krasnoyarsk. *"In the long run, something has to be done about the emissions of GHG."* High temperatures are melting Russia's *"17 million square miles of permafrost above the Arctic Circle"* which caused at least one disastrous industrial accident and threatens the infrastructure, including pipelines, roads, and housing. *"The near-complete disappearance of sea ice off Russia's northern coast this year, 2019",* has proved an economic boon, with shipping companies predicting that year-round navigation through the once icebound Northeast Passage might soon become possible. *"These phenomena are unprecedented. We see changes in atmospheric patterns which leads to more extreme weather events with higher frequencies and duration.* International researchers have concluded that Siberian heat is a result of man-made climate change. In the northern Russian province of Siberia, more than 7 million acres have been burned. The Russian government has been slow to accept this observation, but most Russian scientists believe that what happened in *central Siberia is abnormal in April with temperatures above 30 C [86 F]. The scientists warn about this for at least two decades, and successive Russian governments have failed to take any actions, and they are still in denial".* Melting permafrost caused fuel tanks to rupture at a power station next to the city of Norilsk in Siberia. The spill released 20,000 tons of diesel fuel into the soil and nearby rivers, creating an *"environmental catastrophe. President Vladimir Putin declared a state of emergency in Norilsk".* Lack of official responsibility for the past 50 years, which is to *"ramp up the extraction of fossil fuels, mostly for export."* Like in Brazil in Russian *"the $$ speak and dominate."*

"How is the arctic nearing a point of no return (reached the tipping point)?

With the basic chemistry and physics of how the atmosphere absorbs heat, there's no way that the Arctic is going to cool off again. Where does the source of cold air come from? The energy balance of the Earth is disrupted and warms up. In Alaska, 2.4 million acres of wildfire have burned until the middle of 2019. The largest threat to Greenland is a heatwave blowing from Europe to the Arctic, causing 197 billion tons of ice melt in July 2019. The year 2020 weather events and temperature rise, *"showed that the point of no return has been reached."* The WMO, citing data from the Copernicus Earth Observation Program, says July is becoming the hottest month in recorded history. *"We have this unexpectedly large release of methane from permafrost, then we're going to have to change our assumptions about how fast warming is going to occur, and that change would be faster."* If the Greenland ice sheet melted completely, the global sea level would rise by roughly 7 meters, or 23 feet, according to the National Snow and Ice Data Center. *"What's special about the Arctic is that there is an on and off switch. Once the ice is gone, it changes things dramatically, and that's the process that we're starting to see play out."* (40).

2.2.3. HURRICANES, CYCLONES, AND TYPHOONS

General Notes

In the Atlantic, the big storms are called Hurricanes, in the Indian Ocean and part of the Atlantic, are called Cyclones, and in the Pacific ocean are called Typhoons. In 2020, the Atlantic hurricane season hurled one shocking event after another. What follows is a recap of the most astonishing storms in this unprecedented season. *Atlantic hurricane season 2029, swept 30 tropical storms".* A season only produces 12 named storms. *"North Atlantic sea surface temperatures were, and are far above normal in the entire basin, 2 F (1.1C)".* The West African monsoon was very active this summer. *"The western Caribbean is already the home to the highest ocean heat content in this part of the world". "The warmer the water, the stronger a storm can get."* (WMO).

Seven Category 5 Storms.

Hurricane Iota attained winds of 160 mp/h (256km/h), becoming a category 5 hurricane. It is the most intense storm on record that late in an Atlantic hurricane season. In the last five hurricane seasons, we have seen at least one Category 5 hurricane (1 at each 4.6 years apart), in the last 170 years, the National Hurricane Center estimates there have only been 37 Category 5 storms, (1 at every 4.6 years apart). *"For the 1ˢᵗ time on record (since 1851), we have 5 straight years with a Category 5 hurricane in the Atlantic".* 2016: Matthew, 2017: Irma/Maria, 2018: Michael, 2019: Dorian/Lorenzo, and now in 2020, Iota. For every 2F (1.1C) increase in ocean warming, a storm can gain almost 20 mph in peak wind speed. In a paper published earlier in 2020, NOAA's Dr. James Kossin found major hurricanes (Category 3, 4, and 5) are indeed increasing. Kossin told CBS News, *"Globally, there's about a 25% greater chance now that a hurricane will be a major hurricane intensity than four decades ago. In the Atlantic, there's about twice the chance."* (NOAA). In fact, *"85% of the damage from all hurricanes is caused by major hurricanes".* Stronger hurricanes imply exponentially more damage (41 & NOAA).

"So far this season there have been 12 U.S. landfalls, more than the old record of nine set in 2016". Nine of this year's landfalls occurred along the U.S. Gulf Coast and five of them in Louisiana. Eta becomes the 12ᵗʰ named storm, and first in Florida in 2020. Dual hurricane landfalls occurred just miles from each other on three occasions: Laura and Delta in Louisiana, Delta and Zeta in Mexico's Yucatan, and Eta and Iota in Nicaragua. What are the chances? 3 instances of dual landfalls in one season all within miles

of each other. The most recent is Eta & Iota ~15 miles apart in Nicaragua, Delta & Zeta ~40 miles near Cozumel, and Laura & Delta ~13 miles apart in Louisiana. While many of the records set in 2020 will likely stay in the record books for years to come, one thing is certain: As long as society continues to release heat-trapping greenhouse pollution, climate change will continue to power hurricane seasons that defy the odds (42).

HURRICANES

Dorian, "The Worst Natural Disaster I've Ever Seen."

A Florida-based search and rescue team returned from the Bahamas, describing the devastation sustained during Hurricane Dorian as "*unlike anything they've ever seen.*" Joseph Hillhouse, assistant fire chief at Gainesville Fire Rescue, was part of a six-man team. "*They said we're going where the damage is the worst and the resources are the least. That's how they ended up at Marsh Harbour, which was in the eye of the hurricane on Sept 1,*" On Sept. 1, 2019, "*Dorian made landfall on the Bahamas as a category 5 hurricane, with up to 185 mp/h (296 km/h) winds.*" Catastrophic conditions" were reported in Abaco, with a storm surge of 18 to 23 feet (5.5-7 meters). The storm completely devastated the Abaco Islands and Grand Bahama. According to Prime Minister Hubert Minnis, "*43 people have been confirmed dead, however, 1,500 people are still missing*". In some parts of Abaco, "*you cannot tell the difference as to the beginning of the street versus where the ocean begins,*" Prime Minister Hubert Minnis said. According to the Nassau Guardian, he called it "*probably the saddest and worst day of my life to address the Bahamian people.*" Hillhouse tells TIME that certain areas in Marsh Harbour were so hard hit that there were no remaining buildings. "*It was the worst natural disaster I've ever seen,*" Hillhouse says, placing the devastation at "*somewhere in the 90th percentile. It's an honor and privilege to be able to help in a small way,*" Chief Lane said. "*We're a small team helping a small area that has been hard hit; they've been able to provide critical assistance to an area that was in great need so it's been our honor.*".

After leaving the Bahamas, hurricane Dorian continued its course to Florida. The National Hurricane Center issued a hurricane watch for

Florida's East Coast from Deerfield Beach north to the Georgia state line. The same area was put under a storm surge watch. menaced the entire state.

Hurricane Derecho, Aftermath.

Farmers across Iowa are dealing with the heartbreaking aftermath of a rare wind storm that turned what was looking like a record corn crop into deep losses for many.

The storm, known as a derecho, slammed the Midwest with *"straight-line winds of up to 140"* mp/h (225 km/h) on Monday, August 10, 2020. It *plowed through Iowa farm fields, flattening corn and bursting grain bins still filled with tens of millions of bushels of last year's 2019 harvest."* The damage was estimated at $7.5 billion. *The U.S Agriculture Department has estimated that Iowa farmers will be unable to harvest at least 850,000 acres (343,983 hectares) of crops.* Before the storm hit, the U.S. Department of Agriculture had been expecting a record national corn crop this year of *"15.3 billion bushels harvested from about 84 million acres."* Iowa's crop was valued at about *"$9.81 billion in 2019"* (43a).

Hurricane Laura Will Cause Unsurvivable Storm Surge

Hurricane Laura is expected to cause an *"unsurvivable storm surge"*, extreme winds, and floods as it hits the US, the National Hurricane Center (NHC) reported. *"The category 4 storm is approaching Texas and Louisiana with maximum sustained wind speeds of 150 m/h, (240km/h)."* If it maintains those speeds it would be one of the strongest storms to ever hit the US south coast. *"Half a million residents have been told to leave the area".* Laura and another storm, Marco, earlier hit the Caribbean, killing 24. Marco has already struck Louisiana, bringing strong winds and heavy rain on Monday, August 24, 2020. Initially, it was feared that *"both storms would hit Louisiana as hurricanes within 48 hours of each other, an unprecedented event".* Laura has strengthened rapidly with maximum sustained winds of 140mph (220km/h).

Hurricane Katrina - which devastated New Orleans in 2005, killing more than 1,800 people - was a category 5 storm before weakening to a category 3 when it made landfall in the US.

"Despite the devastation, overall damage not as bad as feared"

Laura killed at least six people with the damage wrought by the Category 4 system is estimated at $14 billion.

"It is clear that we did not sustain and suffer the absolute, catastrophic damage that we thought was likely," Louisiana Gov. John Bel Edwards said. *"But we have sustained a tremendous amount of damage."* He called *"Laura the most powerful hurricane to strike Louisiana, meaning it surpassed even Katrina, which was a Category 3 storm when it hit in 2005".*

After Laura, it was toppled. *"It looks like 1,000 tornadoes went through here. It's just destruction everywhere. There are houses that are gone. They were there yesterday but are now gone. More than 580,000 coastal residents evacuated"* under the shadow of a coronavirus pandemic. *Storm surge wound up in the range of 9 feet to 12 feet (3 to 4 meters),* (still bad, but far from the worst forecast). Laura hit the U.S. after *"killing nearly two dozen people"* on the island of Hispaniola, *"including 20 in Haiti"* and three in the Dominican Republic. It sets a new record for U.S. landfalls by the end of August. The old record was six in 1886 and 1916, according to Colorado State University hurricane researcher Phil Klotzbach. Laura was tied with five other storms for the fifth-most powerful U.S. hurricane, behind the 1935's Labor Day storm, 1969's Camille, 1992's Andrew, and 2004's Charley, Klotzbach said (44).

Hurricane Delta.

Weary residents of coastal Louisiana began cleaning up on Saturday, Oct. 10, 2020, from wind and water damage inflicted by Hurricane Delta to their already storm-battered region (again). Delta made landfall near the town of Creole in Cameron Parish early Friday, Oct. 9, as a Category 2 hurricane with winds of 100 mph, (160km/h). After August Laura, Delta toppled trees and power poles, leaving hundreds of thousands of Gulf Coast residents without power. Delta brought widespread flooding. *"Even if it wasn't quite as powerful as Hurricane Laura, it was much bigger,"* Governor John Bel Edwards told a briefing in Baton Rouge. Some 3,000

National Guard troops had been called up. As Delta made its way over the Gulf of Mexico on Friday, Oct. 9, 2020, energy companies cut back U.S. oil production by about *"92%, or 1.7 million barrels per day"*, the most since 2005's Hurricane Katrina. Delta was the tenth named storm of the Atlantic hurricane season to make U.S. landfall this year (45). The residents of Louisiana, after being buttered several times in a time frame of 3 months by hurricanes and floods, are thinking twice if they stay or move out of the area. The same issues are facing people from hard-hit zones caused by firenado in California, and flooding in different places of the planet, like China, Filipino, and Indonesia. This phenomenon belongs to *"climate migration"*.

Central America Devastation Storms 2020

Villagers in Guatemala's Mayan hillside had been about to harvest their *"cardamom crops that take three years to grow"* when waves of floodwater triggered by two tropical storms last Nov. 2020, washed them away. So, after 3 years of hard work, you lose everything in 2 weeks. Now they have no way to support themselves or to build back the 25 homes, a third of the village was also destroyed in the flash floods. *"No one had ever seen flooding like it around here. The school is flooded, the cemetery is flooded."* And thanks to climate change, *"Central America will have to brace for stronger storm impacts in the future.* Two of the year's strongest storms, Eta and Iota, ravaged swathes of Panama, Costa Rica, Honduras, Nicaragua, El Salvador, Guatemala, and Belize in unusually quick succession in November 2020. More than 200 people were killed and more than half a million displaced. Hundreds of thousands are now unsure where their next meals will come from. But elsewhere in the mountainous central Guatemalan region of Alta Verapaz, storm-triggered *"landslides buried dozens of houses with people inside"*. Hurricane Eta alone caused up to $5.5 billion in damage in Central America, the Inter-American Development Bank said. Nicaragua estimates the damage of both storms at more than $740 million, around 6.2% of gross domestic product (GDP). The new hurricanes are sometimes moving more slowly, stalling for longer on land or traveling farther before breaking up. This kind of behavior will be the *"new norm"*. Hurricane Harvey turned Houston's highways into tidal rivers after stalling for four days near or over the Texas coast, pouring 50 inches (1.25 meters) of rain. *Iota spun from a 70 mile-per-hour (113 km-per-hour) tropical storm to a 160-mph (257-kph) Category 5 hurricane in 36 hours.*

Guatemala's President Alejandro Giammattei said: Central America had been the worst affected region in the world by climate change and it would need help from them to stave off mass migration. (46). These are the new characteristics of the newly developed hurricanes, due to increased global warming. Following are details of 3 storms that hit Central America in November 2020.

Iota Hurricane "One More Hour And We'd All Be Dead"

Colombian island (Isla de Providencia, a remote Caribbean paradise off Nicaragua) smashed by record-breaking category 5 hurricane. Presenting 2 almost entirely overlapping storms in Central America, a rare occurrence. 80% of all structures on Isla de Providencia, no longer exist. The church building's roof was ripped off by the record-breaking 160mph (256km/h) storm. With little left, families of the island will likely be joined by thousands of others made destitute, and move (migrate) somewhere else. Hurricane Iota broke records *"as the largest hurricane ever to hit the area, and also the strongest storm ever recorded. "Iota is the 30th named storm in the Atlantic this year."* Iota later barreled through Nicaragua and Honduras, killing over 40 people, including one person in Providencia. China and the United States have promised hundreds of thousands of dollars in relief aid. *Iota came ashore just 15 miles (25 kilometers) south of where Hurricane Eta made landfall on Nov. 3. Eta's torrential rains saturated the soil in the region, leaving it prone to new landslides and floods, and that the storm surge could reach 15 to 20 feet (4.5 to 6 meters) above normal tides".* The hurricane significantly damaged all structures on the island *"with up to 80% of buildings destroyed."* The 160mph, (256km/h) winds tossed fishing boats into the streets and left piles of rubble where many homes once stood. *"Piles of debris and downed trees still block many of the mountainous island's streets."*

Hurricane Eta, which passed by the island *"just three days before Iota,"* damaged over 50 homes. *"Eta went on to cause significant damage in Nicaragua, Honduras, and Guatemala and killed nearly 200."* Iota marked only the second category 5 storm ever recorded in November. The Atlantic hurricane season ends on November 30 (46a).

Hurricane Sally

"Hurricane Sally has brought "historic and catastrophic flooding" to the southern US after making landfall." The storm's sluggish speed, roughly 5mph (7km/h), increases its capacity for *"destruction, pummelling coastal states with heavy rain"*. The National Hurricane Center (NHC) reported flooding from Tallahassee, Florida to Mobile Bay in Alabama. The NHS said *"historic and catastrophic flooding, including widespread moderate to major river flooding, is unfolding"* in Alabama and Florida, where some areas have already received more than 24 inches of rain, (60 cm). It also warned of a *"life-threatening"* storm surge and river flooding inland as far as Georgia. *"Alabama, Florida and Mississippi have all declared states of emergency"*. Louisiana Governor John Bel Edwards, whose state is still recovering from Hurricane Laura last month tweeted that residents should all have a *"game plan in place for whatever the rest of hurricane season has in store"* (47).

CYCLONES

A shocking image from space shows *"a record 5 tropical cyclones in the Atlantic basin at the same time"*. This includes the Atlantic Ocean, the Caribbean, and the Gulf of Mexico.

The cyclones tie yet another record in an extremely active hurricane season.

The NOAA's Geostationary Operational Environmental Satellite (GOES) captured the image.

It shows Tropical Depression Rene, Tropical Storms Teddy and Vicky, and Hurricanes, Sally and Paulette. *"This ties the record for the most number of tropical cyclones in that basin at one time,"* the National Weather Service tweeted on Monday, Sept. 14, 2020. The Atlantic Ocean, Caribbean, and the Gulf of Mexico have only held this many storms once before, in September 1971. The satellite images also show smoke swirling towards the East Coast from the unprecedented wildfires in California, Oregon, and Washington (NOAA & BBC).

Cyclone Amphan

The storm Amphan is expected to make landfall on Wednesday, May 20, 2020, and hit West Bengal and Orissa (Odisha) states. Both states are seeing large numbers of people return from villages, many are on foot. This would be the *"first super cyclonic storm in the Bay of Bengal since the 1999 super cyclone"* that hit the Orissa coast and killed more than 9,000 people, according to BBC Weather. Heavy rainfall warnings have been issued for West Bengal and Orissa, 200-250mm of rain is expected. Orissa is planning to *"evacuate more than a million people"* from coastal areas.

Hundreds of thousands of people had already been evacuated as the region braced for Cyclone Vayu, which was classified as *"very severe"*. And in May last year, India evacuated more than a million people to safety to avoid Cyclone Fani, in which 16 people were killed in Orissa. In the Bay of Bengal, the cyclone season typically runs from April to December (47a, BBC).

Cyclone Nivar

The cyclone has made landfall in southern India, triggering torrential downpours in coastal areas of Tamil Nadu. The cyclone was categorized as a *"very severe cyclonic storm"*,

Officials said it reached winds of more than 120km/h (75mp/h) but then weakened into a severe cyclonic storm. Tens of thousands of people from low-lying areas had been evacuated ahead of landfall. According to local media, one woman died when a wall collapsed following heavy rain, which caused severe flooding. The heavy wind had already felled trees and strong rains flooded parts of Tamil Nadu and the capital Chennai, on Nov. 26, 2020. (BBC).

TYPHOONS

Typhoon Goni. The Philippines was hit by this year's most powerful storm Goni on Nov. 1st, 2020. At least 25 people were killed, and about 13,000 shanties and houses were damaged or swept away in the eastern island province. Goni blasted into Catanduanes province, with winds of 225 km/h (140 m/h), and gusts of 280 kph (174 mph). The typhoon shifted direction to spare the capital, Manila, before blowing out into the South China Sea. Catanduanes Gov. Joseph Cua said the typhoon whipped up 5-meter (16-foot) storm surges and damaged or swept away about 13,000 houses. More than 66,000 other houses and huts were damaged elsewhere in the region, officials said. 19 million people have been affected by the path of Goni, (more than Romania's population of approx. 18 million people). *"This 19 million already includes the populations in danger zones for landslides, flooding, storm surges, and even a lava flow,"*

Goni, known as Rolly in the Philippines, *"is the most powerful storm to hit the country since Typhoon Haiyan killed more than 6,000 people in 2013."* The BBC's Howard Johnson in Manila says. *"The winds are fierce and very strong."* Goni is the 18th typhoon to strike the Philippines this year, according to The New York Times. The Philippines is used to powerful storms, it is hit by an average of *"20 storms and typhoons a year."* and lost 22 people when Typhoon Molave barrelled through the same region last week. But this year preparations have been complicated by the Covid-19 virus. Crisis on top of the crisis, *"what's life in the 21st Century on the planet Earth."*? Some 347,000 people were evacuated (BBC). Coronavirus patients being treated in isolation tents had been evacuated. *"Evacuating people is more difficult at this time because of Covid-19, the pandemic had depleted their funds for disaster emergencies. People affected by Typhoon Goni were still reeling from the impacts of three previous cyclones that came in October. The Red Cross is ensuring that their urgent needs are supported amid the challenges posed by the Covid-19 pandemic."* (BBC).

Typhoon Vamco (Ulysses) Nov. 12, 2020.

At least 7 people have been killed by a powerful typhoon that struck the Philippines, unleashing some of the worst floods in years in the capital Manila.

Typhoon Vanco, also known as Ulysses, is the eighth to hit the Philippines in the past two months. The President promised that the public should *"rest assured, the government will not leave anybody behind. A lot of places are submerged. Many people are crying for help,"* Rouel Santos, 53, a retired disaster officer in Rizal province told Reuters. Roughly 40,000 homes had either been fully or partially submerged in the Marikina area. The typhoon struck areas that were still recovering from Goni, which killed 25 and left thousands in temporary shelters. Almost 200,000 people were evacuated before Vamco struck with gusts of up to 158 mph. The President, Duterte, said that the devastating storms were *"a stark reminder of the urgency of collective action to combat the effects of climate change"* (48).

A powerful typhoon Haishen damaged buildings, flooded roads, and knocked out power to thousands of homes in South Korea, on Sept. 7, 2020, after battering islands in southern Japan, killing one person and injuring dozens of others.

Typhoon Maysak.

A powerful typhoon, Maysak, ripped through South Korea's southern and eastern coasts with tree-snapping winds and flooding rains and leaving one person dead. Japan's coast guard was searching for a *"livestock ship carrying 42 crew members and 5,800 cows that made a distress call off a southern Japanese island"* in rough seas. A Filipino crew member rescued said the *ship capsized before sinking,* according to the coast guard. During its life as a typhoon, Maysak packed maximum winds of 90 mph (144kph). Officials have managed to restore electricity to about 199,400 of the 278,600 homes that lost power (48a).

The Molave Typhoon is *"one of the two most powerful storms Vietnam has had in the past 20 years,"* according to a statement on the government website citing Deputy Prime Minister Trinh Dinh. The storm to hit

Southeast Asia has left at least "*26 fishermen missing at sea* "Wednesday, Oct. 28. Vietnam is enchanted by nine major storms so far this year, 2020. The storms have resulted in over 130 killed in just its central region. Some 250,000 soldiers are on standby to respond to Molave. In the Vietnam typhoon, almost 50 people are missing and 45 were injured. About 1.3 million people left their homes, as the storm damaged more than 91,000 homes. An estimated 2.7 trillion dong ($116 million) worth of property and crops were damaged by storms and other natural calamities that hit Vietnam in October. The estimate of damage is up to: "*111,900 houses, 45,000 hectares of paddy rice, 22,300 hectares of fruits and vegetables.* "*Molave killed at least 16 people*" in the Philippines (49).

2.2.4. FLOODINGS

General Notes.

All storms presented above are coming with huge surges, rain, and causing devastating floods. In this paragraph, I will try to present the floods which are not covered in storms.

Flooding which occurs in America once in a lifetime could become a daily occurrence along the coastline if sea level rise is not curbed, on more than 90% of the coast. Around mid-century flooding events will become annual occurrences for more than 70% of the US coast. This comes with multibillion-dollar damages and will become uninhabitable. For 5-10cm (2 to 4 inch) of sea-level rise "*can double the chances of extreme flooding every 10 years to the end of the century.*" By 2100, the sea level will be 12 inches (30 cm) since 1880. Almost all rises will depend on the melting, in Greenland and Antarctica (50). Coastal flooding linked to climate change could cost trillions of dollars and affect hundreds of millions of people in the U.S. and around the world by the end of the century. Those floodings could threaten "*assets worth up to $14.2 trillion worldwide, equal to 20% of global GDP. To avoid it, reduce the emission of GHG. "Climate variation increases the frequency, and intensity of weather events, which will further increase the risk of flooding.*"The areas impacted by flooding are portions of the eastern U.S, northwest Europe, southeast and east Asia, and northern Australia (51).

River flooding could surge to as many as 50 million a year by the end of t2100. Half of it will be by increased populations. If society takes measures to reduce carbon emission, *20 million people a year* can be saved from being out of their homes. India, Bangladesh, and China can move millions of people out of harm's way ahead of storms, and this has to be followed by sub-Saharan Africa (52). I will continue with descriptions of floods in different areas (regions) of the planet, which are not part of the floods due to hurricanes, cyclones, or typhoons. In the last book (6), in the Climate Change chapter, I present the so-called *"River Clouds"* atmospheric phenomena, which may develop in California. Will be out of mind to see flooding caused by a River Cloud, of the dimension: 1000 miles long and 300 miles wide. In the history of California, the *"Mega Flood"* disasters occurred many times in the last 1800 years, on an average of 200 years, the last one was in the 1861 and 1862 periods. It is so-called:

"California's Trillion-Dollar' Mega Disaster no One is Talking About"

Disasters on the West Coast include devastating earthquakes and out-of-control wildfires, but there's an epic disaster that could be far worse than both, and it could happen at any time. Officials and experts call it the *"ARkStorm."* Imagine a month of drenching storms along the entire West Coast. *"The state of CA. would be swallowed in 10 to 20 feet (3.05 to 6.1 meters) of rain"*. At up to 200 inches (508 cm) in some places. Any populated area will be flooded. Most populated areas of the San Francisco Bay area, Los Angeles, San Diego, and Sacramento would all be swallowed by water. You can see *"thousands of landslides, major dams, and agriculture industry failures.* This type of storm happened in 1861 and 1862 and is possible now.

"The deadly storms of 1861 and 1862 fundamentally changed California. A quarter of the homes have been flooded, destroyed one-third of the taxable land, and it bankrupted the state."

A sea of water *in the Central Valley of 40 miles wide and 150 miles long has been created"* UCLA Climate Scientist Daniel Swain said. He believes approx. 1% of the state's 400,000 people died. *This is more than a trillion dollar-type disaster with long-term consequences, 1.5 million displaced"*. Emergency Services of CA. has developed a *"mega-storms plan."*

In the last 1800 years, tree rings and rocks show six "mega-storms" more severe than the 1861-1862 storm. Nowadays 1% of the population of CA. to vanish means, 395,000 people, the most catastrophic disaster in U.S. history. California grows the majority of our country's crops, the impact would be felt across the entire country (52).

Tennessee Flooding (Feb. 17, 2020).

Authorities managing dams in Tennessee and Mississippi must make difficult decisions as floodwaters swell along the states' rivers; the surging water pressing against the dams has to be released at some point, and when it does, it often spells disaster for individuals living downstream from the dams. Case in point: two large homes slid into the flood-swelled waters of the Tennessee River. Luckily no one was hurt in the landslides. *"It kills you, knowing that,"*

February rains have been *"400 percent of normal"* and that more was on the way. *"It's kind of a never-ending battle".* Mississippi Governor Tate Reeves called the floods *"historic"* and *"unprecedented."* (53).

Michigan Hit With '500-Year' Flooding

Rising floodwaters as high as five feet (1.52 meters), submerged parts of central Michigan on Wednesday, May 20. 2020, after days of heavy rain, led to the failure of two dams. The National Weather Service (NWS) warned of *"life-threatening"* flooding as water levels reached historic levels. *"Never in my whole life have we seen the dam fail,"* said Mark Bone, resident of Midland. *"It flooded real bad in '86, but never like this."*

Michigan Governor Gretchen Whitmer called on the *"federal government to provide help"* to the state. About 10,000 people have been evacuated in Midland County, heavy rain caused a swollen river to overflow its banks and breach the Edenville and Sanford dams. *"This can be called a 500-year event."* Floodwaters had reached five feet (1.54 meters) in parts of downtown Midland. This flooding is on top of the pandemic crisis (54).

Coastal Flooding in The US Caused by The Seas Rise.

"*Sunny day*" high-tide flooding is becoming more commonplace in the U.S., and a federal report released Tuesday, July 14, 2020, by the National Oceanic and Atmospheric Administration (NOAA) "*warns that such flooding will worsen in the decades to come as seas continue to rise. America's coastal communities and their economies are suffering from the effects of high-tide flooding, and it's only going to increase in the future,*" said Nicole LeBoeuf, acting director of NOAA's National Ocean Service. In 2019 alone, 19 locations along the east coast and Gulf coast set or tied records of high tide flooding. In 2019, the Southeast saw a threefold increase in flooding days compared to 2000. Along the western Gulf coast, percentage increases were greater than fivefold. In Texas, Sabine Pass and Corpus Christi had 21 and 18 flooding days in 2019, and in 2000 one and three days, respectively. By 2030, there will be 15 days of high tide flooding for coastal communities nationally. By 2050, it rises to 25 to 75 days (55).

China Floods. Dam Collapse.

China's dam in Guangxi province, reinforced 25 years ago, was overrun by heavy flooding in June 2020. It caused immense damage. "*Its collapse suggested that neglect left this reservoir open to disaster*". But perhaps "*most concerning of all, is what that might mean for the 94,000 other aging dams across China*". When it collapsed, a Reuter visit to the reservoir found that the 100-meter length of the dam had largely vanished. "*Every dam here has a 50-year history. If you don't maintain them, they won't survive.*"

A Chinese official said in "*2006 that dykes had collapsed reservoirs at over 3,000 over the past 50 years,*" due to substandard quality and poor management (56).

Severe Floods Engulf Eastern China.

In July 2020, floods overwhelmed parts of central and eastern China. Dams across the country's largest freshwater lake, Poyang, overflowed, with the lake reaching record water levels. The authorities said "*33 rivers* had reached record highs", while downpours are continuing to batter regions along the Yangtze River.

"However, the water level of 98 rivers nationwide is still at an alarming level", which maintains a level-three emergency flood response. *"Some 140 people are reported to have died and millions have been evacuated"*. Thousands of soldiers have also been dispatched to shore up the lake's banks in Lushan. (57).

Hundreds of Rivers Swell in China Summer Floods (July, 13/2020)

Floods over central and eastern China that have left more than 141 people dead or missing are hitting record levels. The coronavirus ground zero of Wuhan, through which the powerful Yangtze River runs, is an area watching the rising waters. *"The river is hitting its third-highest levels in history in the city of 11 million."* Steady rains flooded huge areas, leaving 219 people dead or missing, affecting 37.89 million others and destroying 54,000 homes, economic losses of 178.9 billion yuan *"($25.7 billion).* *"Thirty-three rivers have reached record highs, while alerts have been issued on a total of 433 rivers"*. The water level reached the rooftop of the one-story building. China's *"worst floods in recent decades came in 1998 during an El Nino weather effect, killing more than 4,000 people"*, mostly around the Yangtze (58).

The Big China Disaster, "Three Gorges Dam" on Yangtze River.

The problem isn't that China lacks water management projects. *"It has built hundreds of thousands of levees, dikes, reservoirs, and dams on its seven major river systems"*. But many are struggling to cope with months of rain-fed flooding that has ravaged vast swathes of industrial and agricultural land and engulfed millions of homes. *"The Three Gorges Dam on the mighty Yangtze was peaking and could overflow. China has experienced three of the world's 10 most devastating floods since 1950."* Yet flooding in cities is getting worse. *"Annual average of river inundations produces the highest damage in the world."*

The Yangtze River Economic Belt is home to more than 600 million people, accounts for almost 50% of export value, and 45% of GDP. As the water level behind the massive dam rose more than 15 meters (50 feet) above flood level, three gates needed to be opened. 15 meters above the flood level is a catastrophic amount of water. American scientists warned the Chinese government of the worst flooding in the future. (59).

India, Assam State: Monsoon Floods. (July, 15,2020)

"Swathes of the Kaziranga National Park, a Unesco World Heritage site", had been submerged, and that at least 51 wild animals had died. 102 animals were rescued, as some tigers and rhinos strayed into nearby villages to avoid flooding. Kaziranga is home to two-thirds of the world's population of the one-horned rhinoceros (BBC). On July 28, 2020, floods caused by heavy monsoon rains in two of India's poorest states (Assam and Bihar) have displaced or affected 8 million people and killed 111 since May, at a time when coronavirus cases have swelled there. At least 50 people have died in the seasonal floods in India.

On June 1, 2020, starting the monsoon season *"Assam has received 15% more rainfall than a 50-year average and Bihar 47% more".* The floods in Assam, where at least nine one-horned rhinos have drowned, have affected 5.7 million people. More than 45,000 are sheltered in makeshift relief centers. Also, in Bihar, floods have stranded more than 2.4 million people, on top of the difficulty dealing with Coronavirus (Reuter).

Other big floods on the Nile River in Sudan and Ethiopia, in July 2020. In late August Nile had risen approx. 17 meters, the highest in almost a century. Sudan lost 100 people, and 100,000 homes collapsed. In North Darfur, 15 people died and a further 23 are missing (60).

Floods were encountered in many other places in the world in 2020, like Afghanistan, Europe, Africa, and South America. For instance in Romania, thunderstorms, which come with a rain of 0.5 meters (20 inches), pouring in less than half an hour, were causing drastic destruction to crops, livestock, buildings, and roads. Local people are calling this rain in a short time pouring *"Cloud Braking ".* The main characteristic of this type of thunderstorm is that they happen in isolated areas like on top of one village or half the city.

2.2.5. OCEAN CONVEYOR BELT AND POLAR VORTEX.

A major current system in the Atlantic Ocean, which plays a vital role in redistributing heat throughout our planet's climate system, is now moving 15% more slowly than in the last 1,600 years. Part of this slowing is directly related to the warming climate. The Gulf Stream along the U.S. East Coast is part of this system, which is known as the *"Atlantic Meridional Overturning Circulation"*, or AMOC. The weakening of the current is *"unprecedented in the past millennium."* By 2100 the scientists estimate that the circulation may slow by 34% to 45%. That will be enough to reach several tipping points. From the tropics, heat moves northward and sends cold south from the poles. The majority of that heat is redistributed by the atmosphere.

The other part is moved by the oceans in what is called the Global Ocean Conveyor Belt, a worldwide system of currents connecting the world's oceans, moving in all different directions horizontally and vertically. The Atlantic currents move 100 times the flow of the Amazon river.

The current speed moving is influenced by how much ice shelves are melting, producing nonsalty lighter water. *"The current has been relatively stable until the late 19th century."* During the Little Ice Age, 1400s to 1800s was a change in ocean circulation. Wasa a freezing time for Europe. After that, the ocean currents declined, especially in the mid-20 century. (NASA & NOAA).

Polar Vortex

The bitter chill across the U.S. this winter is a sign of what is to come. *"The large, persistent, southward dip in the jet stream responsible for this cold invasion is likely to happen more frequently in a warming climate, as are the warmer-than-normal spells that sit alongside this dip.*

The disastrous impact of the vortex storm resulted in over 120 million

Americans suffering complete cold weather, and *"caused one of the largest power blackouts in American history,"*

In Texas were at least 57 direct fatalities and 13 indirect fatalities attributed to the storm. Blackouts causing at least 4.3 million Texas residents to be left without electricity. The grid was reportedly *"barreling toward a collapse, it was 4 minutes, 37 seconds away,* and was likely *"to plunge the state into darkness for months".* In Dallas it was –2°F, (the water starts to freeze at 32F) the city's coldest temperature since 1930. Houston and San Antonio 13°F. In total, *"356 generators were forced offline by the freeze."* Vortex led to *"disruptions of 332 water distribution systems in 110 counties across Texas".* Power disruptions impacted water treatment plants that forced cities of *"Houston, San Antonio, Fort Worth, Abilene, Killeen, and Arlington, to enact health emergency boil-water orders".* Also, water pipe bursts caused significant damage to numerous residences, business establishments, schools, and health care facilities. Winter Storm Uri could end up *"costing $195 billion to $295 billion."* The storm also *"destroyed about half of the state's citrus harvest, and large agriculture sector including the salad bowl".* State-wide insurance and reinsurance market losses of approximately $19 billion."

In short words, this was the Texas Vortex 2021, with millions of people living in freezing temperatures with no: heat, food, water, lights, during nighttime, transportation, TV, radio, internet for at least a week. The final number of casualties rose to 111 deaths (60a).

3

Fᴏssɪʟ Fᴜᴇʟs, As ᴏꜰ 2ᴏ19, Pᴏʟʟᴜᴛɪᴏɴ

3.1. FOSSIL FUEL EMITTERS

General Notes. There is a huge disparity between the *fossil fuel carbon emitters of* developed countries and those of developing countries. The 37 member countries of the *"Organisation for Economic Co-operation and Development (OECD)"* are generally regarded as developed countries. GHG emissions in these countries have been in decline for over two decades. OECD countries phased out coal, but the non-OECD countries are currently using coal, increasing GHG emissions. In June 2020 BP released its Statistical Review of World Energy 2020. The Review covers energy data through 2019. *"Oil consumption is on top with 33% of all energy."* The next fossil fuels consumption is: *"coal (27%), natural gas (24%).* Green energy: *hydropower (6%), renewables (5%), and nuclear power (4%). Fossil fuels are in the top with 84% of the world's primary sources of energy consumption in 2019"*. China experienced 75% of the world's energy consumption growth, followed by India and Indonesia. Emissions growth in 2019 was 0.5%. Since the 1997 Kyoto Protocol to curb emissions, global GHG emissions have risen by 50%. Oil consumption grew to a new record in 2019. Natural gas consumption rose by 2% in 2019. The share of natural gas in primary energy consumption rose to 24.2%. Yet, *"in 2017, European Union countries subsidize fossil fuels with US$87 billion.* This increase in wasteful and environmentally devastating spending has been driven by a significant shift towards dirtier economic and environmental policies in nearly every EU country. After 2017, the European Union has changed its course pleading toward net-zero emission by 2050 (BP Review Policy, 2020).

3.1.1. OIL AS OF 2019

In 2019 consumption grew by 0.9 million barrels per day (mb/d), or 0.9%, reaching a max. of 100mb/d. China led the growth (680,000. b/d), while the demand for OPEC failed 290,000. b/d. Global oil production fell by 60,000 b/d. The *"value of oil on the market was minus $26/barrel April 2020. In other words, please take my oil and I will pay you for your kindness $26/ barrel."* This never happened in the history of the oil industry in the world. JP Morgan called *the "current crisis unprecedented. This crisis will change the industry in ways no other crisis has done."* One specific aspect of the oil industry before the pandemic crisis was the rise of exploration of new fields of oil. But, during the first year of the pandemic (2020), as

Wood Mac's analysts note, *"upstream value in Africa is down one-third (US$200 billion)."*

Why is Africa a good example? Because it was, before the crisis, the hottest new spot for oil and gas investments. As time shows, as of the middle of 2021, the fossil fuels industry is recovering to the previous level, due to increased demand in China and developing countries. Following is an analysis of big oil producers per region (Mac Kenzie Analysts)

RUSSIA

Russian oil giant Rosneft on November 25, 2020, announced the start of operations for its giant Vostok oil project in the *Arctic,* plans that have been criticized by environmentalists.

Vladimir Putin, Russian president approved a law facilitating Russian investments in the Arctic. The design work for a 770-kilometer (480-mile) oil pipeline and a port had been completed. *"Mineral resources are determining Russia's role in the world. The complete total investment of 10,000 billion rubles ($111 billion)"*, including *"two airports and 15 the industry towns. Estimated reserves of around 5 billion tonnes of oil."* The Economy Ministry forecast in March 2020 that *"oil consumption will peak around 2045"*. In Russia, *the energy sector still makes up 33% of budget revenues,* and renewables are *"less than 1% of the power.* has so far been slow to adapt to the transition. Analysts show that economic growth may be limited to 0.8% a year until 2040 if the country doesn't adapt to the transition to green energy. *"Welcome Russia to the Increased Climate Crisis"* (Russia TASS Agency).

AFRICA started to explore its resources, and is the next *"Oil Field Exploration."* The new field has stagnated during the Covid-16 epidemic, but restarted in June 2021, the world still needs new oil. The **Kavango Basin**, a 6.3 million-acre basin with a potential 12-billion-barrel of oil, was acquired by Canadian explorer Reconnaissance Energy Africa Ltd. (ReconAfrica).

New producing countries: Nigeria and Angola, together with a pump of around 4 million (bpd) of crude and also dominate natural gas. Only a handful of other countries in the region produce more than a couple of hundred thousand barrels of oil and gas equivalent. Now, **Reconnaissance Energy Africa** is looking to Namibia, and Exxon, Shell, and Total SA are hoping to bring Namibia fame offshore. The Kavango Basin has an estimated 12 billion barrels of oil, and 119 **TCF** (Trillion Cubic Feet) of gas, significantly higher than Eagle Ford's 2.4 billion of oil, and 50 TCF of gas (Recon Africa). Egypt has awarded **Chevron** and Shell key exploration blocks in the red-hot Red Sea. The blocks cover around 10,000 sq km and carry an investment of $326 million, Egypt's petroleum ministry said, adding that potential investment would rise to *"several billion dollars"* if discoveries were made (61).

LIBYA. The National Oil Corp (NOC), Libya's state energy company, instructed its operator to resume production in 2020. The field will pump almost 300,000 b/d in Oct. 2020. That would double the overall output in Libya to around 600,000 barrels daily, and reach 1,000,000. b/d by March 2021. To mention that Libya is an OPEC member and is *"Africa's largest crude reserves"*. An alliance, led by Saudi Arabia and Russia, planned to extract 2 million b/d from the start of 2021 (62).

GUYANA. The **Liza field** with resources is estimated at 6 billion barrels of oil equivalent. A second field, Stabroek block, of oil with as much as 220,000 gross b/d, is under construction. This will facilitate the production of more than *"750,000 gross b/d from the Stabroek block by 2025."* Guyana, with 800,000 people, may produce more crude per person than any other country in the world. Guyana may become a QATAR of South America.

The **Norwegian** government has pledged to reduce carbon dioxide emissions in line with the EU bloc. Norway is planning a major expansion of oil exploration in the Arctic. The government *will auction up to 136 new oil exploration blocks, with 125 of those in the Arctic Barents Sea*. Norway, the biggest oil producer in Europe, has the world's biggest sovereign wealth fund, worth over $1 trillion, on the back of its oil wealth. This shows that *"Norway has failed to take the climate crisis seriously"* (Bloomberg).

CHINA. Tensions surrounding the South China Sea have always been largely about oil and natural gas. Is there much oil and gas waiting in the South China Sea? Estimates vary wildly. But more important is the waterway, including the resources below the seabed. China is importing crude oil, besides what they produce, and the consumption is high enough to be the biggest emitter of GHG in the world and will peak in fossil fuels consumption, not before 2040.

Iraq Deal. Iraq is poised to sign a multibillion-dollar contract with China ZhenHua (means Revitalize China in Mandarin) Oil Co. The deal is the latest example of China, via state-controlled trading companies and banks. The bidder will buy 4 million barrels a month, or about 130,000 b/d. They will pay upfront for one year of supply, which brings in more than $2 billion, according to Bloomberg calculations (Bloomberg).

INDIA plans to *"nearly double its oil refining capacity"* in the next five years, Prime Minister Narendra Modi said, on Saturday, November 21, 2020.

The country's energy minister was quoted in June 2020 as saying: *"India's oil refining capacity could jump to 450-500 million tonnes in 10 years from the current level of about 250 million tonnes."* India is aiming to *"raise the share of natural gas in its energy-consumption mix by up to four times".* The renewable cleaner energy plans will not reduce fossil fuel consumption, on the contrary, will increase emissions and aggravate the climate crisis (63).

POST COVID-19, Oil Stands.

On Oct. 13, 2020, in IEA's annual World Energy Outlook 2020 report, the OECD energy watchdog states that it doesn't see a peak oil demand before 2040, only a possible oil demand flattening. The report bluntly states that after recovering from the "exceptional ferocity" of the COVID-19 crisis, *"world oil demand will rise from 97.9 million bpd in 2019 to 104.1 million bpd in 2040".* The current assessments are all taking into account a wide range of *"proposed and/or signed energy transition",* and net-zero emission government policies. These will have an impact if fully implemented by all. (There is no such agreement on the International level). Global economies have to come back, and it is a full focus on climate change and renewables. The attention will be focusing on the peak oil demand issues, which will recover quickly. *"Demand will be higher, consumption will reach at least 5-10 million bpd above 2019 levels by 2030, possibly hitting 115-120 million bpd by 2040"* (64 IEA Oct. 2020).

OIL PEAK

The COVID-19 pandemic last year, 2020, has dented oil consumption. While there is no consensus between oil and gas producers on when oil demand could peak, it will create confusion, and fossil fuel companies will take advantage of the situation. A few examples:

BP - Could be as early as 2019. BP forecasts COVID-19 will knock around 3 million b/d off by 2025 and 2 million b/d by 2050. In its two aggressive scenarios. But lately, the assumption is that oil demand peaks at around 2030 - 2035.

Equinor - 2027-2028 Norwegian oil and gas firm Equinor expects global oil demand to peak by around 2027-2028. Equinor sees oil demand at 99.5 million b/d in 2030, and falling to 84 million bpd in 2050, under its central scenario, dubbed Reform. (A 15.5% reduction is far from what needs to be).

Bernstein Energy - 2025-30. Analysts at Bernstein Energy said global oil demand will again stand at 2019 levels of around 100 million bpd by 2023 before soon plateauing. *We expect demand will not peak until sometime in 2025-30."* ???

Rystad Energy - 2028. Norway's biggest independent energy consultancy Rystad Energy sees global oil demand peaking at 102 bpd in 2028.

Goldman Sachs - after 2030. Growth in developing countries will push a peak beyond 2030, *"We do not expect global oil demand to peak before 2030".*

OPEC - Around 2040. In its first-ever prediction for global oil demand to the peak. The Organization for the Petroleum Exporting Countries said demand would recover in the next two years but plateau by 2040.

Shell - no prediction. Royal Dutch Shell has held off on giving any prediction.

"To slow down climate change, these predictions look like death on Earth" (IEA).

Oil Stays for Many Decades to Come.

The oil company executives at CERAWeek by IHS Markit were adamant on Monday, March 1ˢᵗ, 2021, that *"crude demand will rise over the coming decade and that the fossil fuel will remain a crucial part of the energy mix."* Climate change and renewable fuels are taking center stage at this year's gathering of energy leaders, investors, and politicians from around the globe. *"Oil demand may continue to climb over the next decade"* even as countries work to comply with the Paris climate agreement's Hess Corp CEO John Hess.

"We don't think peak oil is around the corner, we see oil demand growth for the next 10 years," said Hess. *"We're not investing enough to grow oil and gas in the future,"* he said.

3.1.2. NATURAL GAS AS OF 2019.

Natural gas *"consumption increased by 78 billion cubic meters (bcm) or 2%,"* below the growth seen in 2018 (5.3%). That led to a global gas primary energy growth to a record high of 24.2%. The increase in gas demand was driven by the USA (27 bcm) and China (24 bcm), while Russia and Japan saw the largest decrease (10 and 8 bcm respectively). Liquid Natural Gas (LNG) supply growth was led by the USA (19 bcm) and Russia (14 bcm). Natural gas is a major contributor to climate change, due to the high volume consumed. That creates a global problem. May be a substitute for fossil natural gas in the form of renewable natural gas (RNG), which is indistinguishable from fossil natural gas. RNG could be made from biomass or captured carbon dioxide. For future extraction of natural gas, let's look at some examples.

SAUDI ARABIA. Aramco To Invest $110 Billion In A Huge Gas Field.

*"Aramco plans to invest $110 billion in the development of the **Jafurah** gas field"* that is estimated to hold some 200 trillion cu ft of gas, (5.66 trillion cubic meters - bcm) the Saudi Press Agency reported. Production begins in 2024. Besides gas, Jafurah could also produce around 425 million cu ft of ethane, and 550,000 bpd of gas liquids and oil condensates. (1 bcm=35.32bcf). The Saudi company is committing billions of dollars to new petrochemical projects, including a US$44-billion refinery and petrochemical complex in India and a US$7-billion investment in another complex in Malaysia (65).

Flaring - Texas.

"Natural gas flaring or deliberately burning gas produced alongside oil," has surged with a crude production in Texas, and is worsening climate change by releasing carbon dioxide, and methane. Texas' flaring intensity is lower than other oil-producing areas, including *"North Dakota, Iran, Iraq, and Russia".* Its flare volumes - around 650,000 thousand cubic feet per day (Mcf/d) in 2018. Texas has issued more than 35,000 flaring permits since 2013 and has not denied any, according to the state commission. *"Neither companies nor regulators have kept up with this challenge,"* said Colin Leyden, a policy advocate for the Environmental Defense Fund, which tracks flaring, adding *"pointing fingers at Iran and Iraq does nothing to fix the problem."*

In the Permian Basin underlying Texas and New Mexico, flaring and venting totaled about 293.2 billion cubic feet in (2019), according to independent energy researcher Rystad *"up about 7% from 2018."* Besides Africa, countries from the American continent, South and Central Asia, Russia (to expand the exploration in Antarctica and Arctic Sea beside its vast Siberia territory), all have plans of *"Extending the Extraction of Gas Oil".* (65a). *"Europe and the USA (Biden Government) are pledging "Net Zero" emission of greenhouse gases by 2050. Or earlier. How will the rest of the World, with these astronomical quantities of oil and gas new exploration, comply with the Paris Agreement?"*

Is Natural Gas Still the Fuel of the Future?

BP now says it will *"reduce its oil and gas production by 40%"* by 2030 as part of its pivot to low-carbon operations. But many will wonder why natural gas, *"the bridge between our fossil fuel present and renewable energy future"* is being included on the chopping block. Despite its much lower emissions footprint, *"natural gas is also under threat not only by the bare fundamentals but by the trends shaping energy policies."*

The **EU Green New Deal** is perhaps the best example of these trends and their effect on future fossil fuel demand, including natural gas. The deal is pretty ambitious and targets *"emission reductions of between 50 and 55% from 1990 levels by 2030 and 100 % by 2050."* According to Wood Mackenzie analysts led by Massimo Di Odoardo. this will result in the *"loss of some 45 billion cubic meters of natural gas demand"*, *"Europe consumes about 500 billion cubic meters of natural gas"*. And Europe's biggest economy, Germany, is determined to complete the Nord Stream 2 pipeline. So it seems *"Europe is preparing for more gas demand, not less"*.(EU Bruxel, Energy Commision)

Asia, the world's biggest gas demand growth. Asia imports *"340 billion cubic meters of gas"* equivalent in LNG, according to Wood Mac, and *"these imports are set to double by 2040"*.

India is a case in point. Gas has a market share of just 6%, well short of the government target of 15%. The biggest growth is the United States, where demand grew by *"27 billion cubic meters"*. In Asia, demand grew by 24 billion cubic meters. There is no immediate danger for global gas demand. Alternative sources of energy are getting increasingly cheaper, and this is not a trend that is about to get reversed (66).

3.1.3. COAL AS OF 2019

Coal consumption declined by 0.6% and its *"share in primary energy fell to the lowest level in 16 years at 27%"*. China and Indonesia had the largest increase in coal consumption.

The OECD consumption fell to its lowest level since 1965. Coal is the most polluting fossil fuel on the planet. In the most polluting country in the World, **China,** the use of coal is increasing. 148GW of heavily polluting, coal-fired plants are either being built or are about to begin construction, according to a report from Global Energy Monitor. The EU's current capacity is 149 GW, the same as what China is adding. Coal power station under construction: *"China 121, India 37, Japan 9. Coal station in operation: China 987, EU 149, UK 12."* What's worse, is that while the rest of the world has been largely reducing coal-powered capacity over the past two years, *"China's new coal power plant offsets the decline elsewhere."* About 121GW of coal power is under construction in China. This figure still dwarfs the pace of new construction elsewhere. *"Only in 2019 China's net additions of coal power were +25.5GW, the rest of the world saw a net decline of -2.8GW.* The pollution created by China affects global warming, not only in China. Pollution does not have borders. *"This is the way of thinking of people with one of the oldest civilizations in the World?"* (66).

Scaling Down Coal Power. Most European countries are closing coal power plants for two reasons: are too pollutant and GHG emitters, and too expensive to run. Wind and solar projects are cheaper to build, according to an analysis released on June 30, 2020. *"Coal had reached a financial "tipping point" making it uncompetitive in most markets".* By 2025, 73% of existing coal power plants will be more costly to run versus renewable energies. that replacing the entire coal plants can be done by 2022. All these disadvantaged aspects of coal, require an urgent transition to green energies. IPCC said that for the 1.5C goal to remain in reach, *"global coal use must decline by 80% below 2010 levels by 2030."* 81% of the European Union's coal plants were uncompetitive today, June 2020, without state support the plants would cease to be a concern. *"Closing coal capacity and replacing it with lower-cost alternatives, but could play a major role in the upcoming economic recovery (IPCC).*

THE UNITED STATES. *"JP Morgan Chase said it plans to phase out loans for coal mining, following similar commitments by Goldman Sachs and BlackRock".* Within the last year of 2019, at least eight major US coal mining companies have declared bankruptcy.

"Coal consumption in the United States dropped a record 18% in 2019, causing a 2.1% reduction in GHG emissions". More than 200 coal plants have closed since the Great Recession. The Rhodium Group said in 2018 that half the U.S. coal power plants could be gone by 2030. Coal's decline

due to competing with abundant natural gas. Natural gas became America's leading electricity source in 2016 with lower emissions than coal *(67)*.

Also, **Murray Energy Corp**, one of the largest privately-held U.S. coal miners, filed for bankruptcy. At 12:09 p.m. local time on Monday, **November 19, 2019,** *the largest coal-fired power plant in the western U.S. permanently closed, will now spend the next three years being dismantled and decommissioned."* Coal accounts for less than *"a quarter"* of the nation's mix. *"Since 2010, power generators have retired more than 500 coal-fired power plants with more than 100 Gigawatts."* A typical nuclear reactor has about 1 Gigawatt of capacity. In the last decades, natural gas has displaced 103 coal-fired plants. At the end of 2010, the U.S. had 316.8 Gigawatts (GW) of coal-fired capacity in the United States. By the end of 2019, coal power decreased by 25% (EIA).

AUSTRIA *"shut down its last coal-fired power plant on April 17, 2020, as part of the country's plan to end the use of fossil fuels by 2030."* to complete its transition to generating energy from renewable sources. Many countries are phasing out or immediately ending the use of coal-fired power plants. Sweden became the third European country to exit coal following Austria and Belgium, which shut down its last coal-fired plant in 2016. Six European countries are expected to follow suit by 2025 or earlier.

The UNITED KINGDOM. The UK's biggest electricity plant is close to using only biomass. Drax Group power plant will complete its switch this year 2021, using organic matter alongside fossil fuel to slash carbon emissions. Biomass has become Britain's second-largest renewable energy source behind wind power, with only a few coal-run plants remaining in the UK. The Drax operation, providing 4 million households with electricity.

"Britain's National Grid will operate a completely carbon-free electricity system by 2025, and would be the world's first". Renewables hit a peak share of 67.5% of electricity in May 2020. *"Four of the plant's six reactors use wood pellets and a carbon-capture system",* while Drax intends on becoming *"carbon negative by 2030",* by removing more CO2 from the atmosphere than it emits, *"ban the use of coal by 2025"* (68).

SWEDEN "*has closed the country's last coal-fired power station two years ahead of schedule, in 2017.* The plant at Värtaverket, in Hjorthagen in eastern Stockholm, described the closure as "*a milestone*" for clean energy in Sweden. The Swedish capital is on track for its district heating to be produced entirely by renewable or recycled energy by 2030. District heating, which is used in many European cities, tends to provide higher efficiencies and less pollution than localized boilers in each dwelling. The company had previously said it would aim to shut down its coal operations in Hjorthagen in 2022 (68a).

GERMANY. German lawmakers have finalized the country's long-awaited phase-out of coal as an energy source, "*shutting down the last coal-fired power plant by 2038*" and spending some 40 billion euros ($45 billion) to help affected regions cope with the transition. As mentioned before, this will be unacceptable for the European Community.

Achieving that goal is made harder than in comparable countries such as France and Britain because of Germany's existing commitment to also phase out nuclear power by the end of 2022. "*Germany, the country that burns the greatest amount of lignite coal worldwide, will burden the next generation with 18 more years of carbon dioxide.*" Chancellor Angela Merkel is accused of making a "*historic mistake,*" saying an end date for coal of 2030 would have sent a strong signal for European and global climate policy (69).

SPAIN. The Spanish "*utility plans to permanently shut its two remaining coal-fired power stations this year 2020*", replacing them with new wind and solar capacity.

Spain was anticipating exiting the fuel by the end of the decade 2030, while the U.K. will shut all its plants by 2025. Spain burned 70% less coal in 2019 than in 2018. The output is poised to fall as Iberdrola and Naturgy Energy Group SA plan to retire their plants this year 2020.

"*Energias de **Portugal** SA and Viesgo Holdco SA will shut their units by 2025.* That would leave Endesa SA, which will be the only remaining operator after 2025. The nation's leader in exiting coal is helped by having some of the best renewable resources in Europe. "*Spain is the most* *oversupplied electricity market in Europe. So it can shut coal power plants by 2027* (70).

ASIA. The UN chief warned Asia to *"quit its addiction to coal."* Several Asian megacities, including Bangkok, Ho Chi Minh City, and Mumbai, are at risk of extreme flooding linked to global warming. **Antonio Guterres** said: Asian countries need to cut reliance on coal to tackle the climate crisis, which he called the *"defining issue of our time"*. *"There is an addiction to coal that we need to overcome because it remains a major threat concerning climate change,"* he told reporters. He said countries in the region need to be on *"the front line"* of the fight by introducing carbon pricing and reforming energy policies. About one-third of Vietnam's energy comes from coal power with a few new plants set to come online by 2050, while Thailand is investing in fossil fuels. New research showed that *"at least 300 million people worldwide"* are living in places at risk of inundation by 2050. (See flooding paragraph with devastating floods and dams failure in China, July 2020 (71).

3.2. EMISSION OF GREENHOUSE GASES

General Notes.

The emission of GHG is in accordance with the consumption of fossil fuels. The most emitter of carbon dioxide in the world in 2019, Megatons CO_2/Yr.

World wide (100%); (39. Gigatons in 2019) World International Aviation (3.5%) World International Shipping (2.9%) The other big emitters, differ from country to country and are presented in paragraph 3.1.

The most emitter in the World per country: China (27%), The USA (11%), European Union (6.4%), India (6.6%), Russia (4,76%), Japan (3.56%), Germany (2.15%), South Korea (1.82%), Iran (1.81%), Saudi Arabia (1.72%), Canada (1.66%), Brazil (1.33%). Reported to the year 1990, only European Union, Italy, and France, which are higher emitters, had a negative percentage of emission: -19%, -16.2%, -12.4%, respectively. (Updated May 16th, 2021). Also, a negative % had Romania, Russia, and Poland, but in these countries, at the end of 1989, their communist governments fell, closing many fossil fuel plants (73).

A new study put *"20 of the world's biggest fossil fuels"* in the spotlight, The Guardian reports. The analysis by Richard Heede at the Climate Accountability Institute (CAI) reveals that the *"20 companies have contributed to 35% of all energy-related carbon emissions since 1965."* Among the firms are Chevron, ExxonMobil, and BP, which contributed the second, fourth, and sixth. Meanwhile, 12 of the 20 companies are state-owned entities, including Saudi - Aramco, Russia's - Gazprom, and the National Iranian Oil Company. Aramco leads the pack by a good amount along with 4% of emissions. *"The Guardian"* reached out to the companies, and several of them replied.

"These companies and their products are substantially responsible for the climate emergency, have collectively delayed national and global action, and can no longer hide behind the smokescreen that consumers are the responsible parties," Heede wrote. Fossil fuel demand was back at pre-pandemic levels in December 2020. *"Shell said in July 2020. "While global CO2 emissions also likely peaked in 2019, (author note: the pick won't happen in this decade 2021-2030). As new data shows, after the pandemic, the consumption of fossil fuels increased with new field extraction.* Other measures such as carbon capture and storage (CCS) and greater use of hydrogen were needed, DNV GL said. *"COVID-19 has shown that behavioral changes are* good for the climate (CAI, DNV GL).

3.2.1. AGRICULTURE

In the year 2000, the UN climate report was more or less a stand-alone warning. Now the world watches as students walk out of classrooms en masse, calling for better climate policies. Narratives like *"An Inconvenient Truth, Drawdown, and Six Degrees"* have made their way into popular discourse. In the past 19 years, policies have become more rigorous, scientific insights have gone deeper, consumer awareness has improved, and we've created major advances in technology. One of the most significant carbon sinks available to us does not require much technology through *"agricultural soil."*

"Regenerative *growing practices",* which avoid tilling and minimize soil erosion and respiration, have the potential to maintain a significant portion of the carbon in the soil while improving the nutrition in our food. US farmers are already practicing this today. In the United States, *"24% of farmers"* have been reportedly using diverse crops already. In 2016, 21% of all cultivated US cropland was subject to no-till farming. For other regenerative practices, an estimated 12% of farms practice residue grazing in the country's corn belt; 8% of US farmers planted cover crops in 2017, and 6% use nitrogen management programs. For maximum impact *"all of these practices should be implemented simultaneously."*

How does it work? Photosynthesis is the operative mechanism here, as plants capture carbon dioxide from the air to build their stems, leaves, and roots, and release the remaining carbon deep within the ground. Carbon-enriched soils have demonstrated greater resilience to floods and droughts. Carbon, up to 15 tons of it per acre per year according to the USDA, can stay locked in the earth. *"If we pair capturing and storing atmospheric CO2 with reducing our emissions, we could have caused real hope of bending the arc of climate change".*

Dirt and money. We do need to create financial incentives for enough farmers to change their practices. *"If farmers provide the societal benefit of removing atmospheric carbon dioxide by adopting regenerative practices, it seems reasonable that they should be compensated for their effort by those of us who benefit, whether we are consumers, corporations, nonprofits, or governments."* Today, the average farmer in the United States makes less than $40 per acre.

"At $15 to $20 per ton of carbon rate, which is lower such as direct carbon capture technologies, estimated to cost between $94 and $232 per ton". If a farmer captures and stores two to three tons of carbon per acre per year, that represents an additional $30 to $60 per acre of bottom-line profit for the farmer. This requires introducing a carbon price. This issue will be discussed in further chapters (74).

The rollout challenge.

If $15 per ton of carbon sequestered growers are willing to transition toward regenerative practices. We can now measure the amount of carbon in the soil with new technology based on satellites, remote sensors, and artificial intelligence. The agriculture sector needs a *"national plan to roll out a regenerative farming practice"*. Educational training and certification for growers remain a hurdle particularly if we look beyond the United States. Especially in American society by definition, climate action *"requires people to scale back on conveniences and pleasures, from disposable plastic bags and straws to air travel to cheeseburgers, and to alter our everyday habits that have in some part helped to so burdened our planet"* (74).

Gardeners should ditch peat compost, as new research shows that *"restoring boglands could offset 33% of all agricultural emissions."* Currently, *"peatland releases 3 million tonnes of CO2e a year, but it could lock up 14 million tonnes of CO2e per year"*, according to a study by the bird charity's scientists. Most of our peatlands are in poor condition and unprotected, however, restoration could lead to higher levels of sequestration too. In a paper published in June 2020 in *Biological Conservation*, Dr. Rob Field, lead author of the paper said: *"Currently our nature-rich lands are already doing an excellent job; they hold a massive store of around 0.5 Gigatonnes of carbon, around 30% of our land-based store on just 20% of its area, as well as capturing an additional 8.7 million tonnes CO2e every year. However, this store and processes are at risk because many of these important habitats are in poor condition, and two-thirds lack any form of protection."* According to environmental experts, *"peatlands are some of our most important conservation areas"*, but most are degraded by drainage, erosion, and inappropriate management, meaning that they are releasing carbon at an alarming rate, losing around 3 million tonnes of CO2e a year. *"The large-scale removal of peat from bogs in Britain and Ireland is destroying one of our most precious wildlife habitats"*. It takes centuries for a peat bog to form - modern machinery destroys it in days. Peatlands provide important nesting and feeding grounds for many wading birds such as dunlin, curlew, and greenshank (75).

Farms Greenhouse Gases.

Agriculture, one of the world's biggest greenhouse gas contributors, is projected to account for about *"20% of total emissions in the next 20 years, mostly methane"* produced by belching cattle and rice paddy fields. And levels are rising as more food is grown to feed an expanding global population. Environmental groups are urging consumers to reduce the amount of *"beef and lamb"* they eat and cutting food waste, which could have the biggest impact on emissions, said the McKinsey report. But farms may cut 4.6 billion tonnes of CO2e a year by 2050 compared with emissions. The first step in reducing emissions is, *"to change how we farm,"* the report said. Of 25 measures highlighted, replacing tractors and combining harvesters that use green energy (hydrogen) would have the biggest impact. *"Of the 19.9 billion tonnes of CO2 equivalent, a year now emitted by agriculture, forestry, and land-use change* (50% of total annual emission), *methane produced by cattle and other ruminant animals contributes about 8.3 billion tonnes"*, said McKinsey. Better fertilization could cut methane emissions from rice paddies, which now emit 2.1 billion tonnes of CO2e. Subsidies could support the greater use of sulfate fertilizer that competes with the bacteria (76). *"We've waited so long to start to address the climate crisis. We will need to both reduce emissions drastically and take as much carbon out of the atmosphere as we possibly can."* Unlike 2006, when Gore's film *"An Inconvenient Truth"* was met with skepticism in some quarters, 13 years of intensifying weather events have persuaded more Americans that climate change is a very big problem. *"Planting trees and sequestering carbon in the soil is likely to remain the two most effective approaches,"* Gore said. The world's population currently uses more than *"33% of the planet's surface for agriculture"*, according to the United Nations. In the U.S., *"close to 40% of the land is farmland"*. The soil used for agriculture has degraded and eroded over centuries of use, losing between 20% and 60% of its original carbon content, according to the IPCC. Researchers show that soil can sequester carbon at rates as high as 2.6 Gigatons each year. An aggressive, global combination of tree planting and increased vegetation along with soil carbon sequestration, has the *"technical potential"* to absorb *157 parts per million of CO2.* With about 415 parts per million in the air today, a huge jump compared with a few decades ago, removing even a fraction of that could slow the advance of global warming. As of 2012, the government said there were 914 million acres of farmland in America. It would cost about *"$57 billion to convert another 1 billion acres by 2050, but that some 23.15Gigatons of CO2 could be sucked out of the atmosphere. This gets to $2.46 per ton of CO2". "For most people, this is a very new idea,"* said Will Rodger, director of policy communications at the American Farm Bureau Federation. *"We're certainly aware that carbon can be stored in the soil, but farmers have very narrow margins. Many of our members are looking very closely at it, but the question is how to make it a business."* (77).

Because humanity has celebrated in the year 2020, the 50th anniversary of Earth Day, I feel responsible to say a few words about this celebration.

Earth Day, 50 years after the first Earth Day.

We are amid a pandemic. And we need science more than ever.

On April 22, 1970, 20 million people marked Earth Day to spread public awareness of the environment. The decade that followed saw some of America's most popular and powerful environmental legislation, according to the Earth Day Network, including updates to the *"Clear Air Act and creation of the Clean Water Act, the Endangered Species Act, and the establishment of the Environmental Protection Agency"* (EPA).

"The pandemic provides unequivocal evidence of the dire consequences of the government ignoring science," marine conservation biologist and environmental activist Rick Steiner said. *"Scientists tell us that as we destroy nature, habitats, and the natural barriers between humans and wildlife we will only see more and more zoonotic diseases like the COVID-19 pandemic,"* Indeed, despite the *"existential threat of climate change,"* today the U.S. and other countries are rolling back environmental protections. John Oppermann, the executive director of the Earth Day Initiative, said *"a disregard for science and basic facts are killing us. It is fueling epidemics and it is driving climate solutions further and further from reality"* (77a).

3.2.2. HOUSING.

If humankind is to make meaningful progress in the fight to limit global warming, we must fundamentally *"alter how we create and consume energy"* in housing, transportation, agriculture, industrial processes, and so on. Unfortunately, much of our global energy infrastructure was built under the presumption that we could continue to burn fossil fuels. So shifting to renewable energy also means shifting some of the systems we rely on to deliver that power. In the USA Democrats are calling for an ambitious $2 to 4 trillion, four-year investment in climate-related projects, including *"upgrading 4 million buildings."* There's also an easy step we could take right now: adopting "***building standards*** *that support a shift to renewable energy sources"*. The International Code Council is a U.S.-based association of government building safety experts. The late last year 2019, the membership voted to significantly update its standards. *Emissions from buildings account for nearly 40% of the greenhouse gases in the U.S* (this percentage varies from country to country). Making buildings more energy-efficient saves consumers money in the long run. And designing new buildings to be ready for renewable power sources, costs much less than it will cost to go back in a few years and retrofit the buildings (78).

WoHo wants to make constructing buildings fast, flexible, and green with reusable components. Housing and building costs continue to rise. *"Why can't the whole process be more flexible and faster?"* Well, a trio of engineers and architects out of MIT and Georgia Tech are exploring that exact question. They start up a new company called *"WoHo (World Home)"*, which's trying to rethink how to construct a modern building by creating more flexible "components" that can be connected to create a structure. The goal of WoHo is to lower construction costs, maximize flexibility for architects, and deliver compelling spaces for end-users, all while *"making projects greener in a climate unfriendly world."* With WoHo, *"it is the integration of the process from the design and concept in architecture through the assembly and construction of that project. Our technology provides the best outcomes for the mid-to-high rise."* It's still early days, but the group has already gotten some traction in the market. The company is building a demonstration project in Madrid and targeting the second project in Boston for next year, 2021 (79).

We can build all-electric buildings today. Cities should require them.

The American Institute of Architects California (AIAC), an association of 11,000 architects in California, wholeheartedly agrees that now is the time to insist that future buildings are designed to be more *"energy-efficient and to be ready for renewable energy sources. We are actively supporting the adoption of an all-electric energy code for residential and commercial buildings"*. We are actively supporting efforts by local governments to require new buildings in their jurisdictions to be all-electric before it becomes a state mandate. All-electric buildings of all types and sizes are already being designed today. We encourage the city of Los Angeles to join other communities in California, by requiring buildings to be all-electric (80).

Living in a house that doesn't fully meet your needs might have been tolerable.

The UK to meet its climate change commitments, *"all houses, old and new, must be zero-carbon by 2050."* Currently, *"houses account for about 28% of all carbon emissions worldwide"*, half of which comes from the energy used for heating and air conditioning. New houses can be built to zero-carbon standards on a cost-competitive basis. For example, between April 2019 and March 2020 in Scotland, 14,000 new homes were built but *"just eight achieved the highest gold level"* rating for energy efficiency. In 2050, we will still be living in about 80% of the homes that exist today, so retrofitting these will also be essential. In January 1988, I published a scientific article in *"Bauphysik"*, a German technical publication (81), laying out the best solutions to improve the thermal insulation of existing buildings. Based on that study, all countries in Europe (with continental weather climate) have taken measures to improve their codes and standards for upgrading building insulation. I want to emphasize that in the construction sector, nations need to include recycling some construction materials, as well as introducing new, improved energy-efficient appliances based on the latest technologies. Following the British saying of *"I am too poor to buy cheap pants,"* we need to change our mindset and produce durable products that can be repaired or upgraded with new parts, not replaced once (6).

In the UK, "*the construction industry contributes about £117 billion to the UK economy annually, of which housing is about £38 billion (33%)*". The sector provides "*more than 2.5 million jobs and apprenticeships, about 7% of total UK employment.*" Britain has a target to reach net-zero emissions before 2050, which will require a huge reduction in the use of fossil fuels, "*such as gas, which is currently used to heat about 85% of the country's homes*". A national drive to "*decarbonize housing can form the basis of good, well-paid work within every community in Britain*". This model can be adapted and used by any country in the world.

The EU Plans a big building renovation project to save energy. EU Commission President Ursula von der Leyen said on Oct. 12, 2020, "*that the 27-nation bloc "must speed up" the pace of renovations if it is to meet climate change targets. "Our buildings are responsible for 40% of our energy consumption,"* said von der Leyen. And even though many buildings have or are being renovated, "*at the current pace, it would take more than a century to bring emissions from our buildings to zero. Some of the financial aid involved is set to come from the 750 billion euro recovery fund backed by the EU.* The European Union has set a target to renovate 3% of government buildings each year, which is far behind meeting climate targets.

An investment for future generations.

Everyone needs somewhere to live, and it's estimated that "*two billion new homes will be needed globally by 2100*". That includes accommodation beyond just houses. We must recognize our housing stock is a long-lasting part of a society's infrastructure, of value to people now and future generations. This is especially true for decarbonization, a modest increase in the cost could deliver significant energy savings throughout its entire life cycle. The post-pandemic recovery "*is both green and socially equitable, addressing two of the greatest crises of our time.*" The same issues with buildings, especially homes, are in all countries, with particularities of each country's culture, and level of development.

Specific Real Estate Market in the USA

The investment firms can profit from *"the lack of attention being paid to the risk of climate change in a property industry that's building, buying, selling and lending (cheaply) without taking into account rising sea levels and inland flood risks"*. The problem starts with government flood maps that under-report risk. Between *$60 billion to $100 billion* worth of mortgages **for U.S. coastal homes** are issued each year. *"Some 311,000 existing coastal homes will be repeatedly flooded, or lost altogether, within the next 30 years"*. That means often high-population states of California, Texas, Florida, Maryland, New Jersey, and New York are vulnerable. In 2016, Freddie Mac's (Real Estate Mega Co. in the USA) then–chief economist Sean Becketti wrote that *"the economic losses and social disruption [of rising seas on coastal housing] may happen gradually, but they are likely to be greater in total than those experienced in the housing crisis and Great Recession. We have no idea how bad things really could get, there are far bigger risks associated with climate change than depreciating home values."* The same situation with *"housing is in different parts of the planet, like Southeast Asia, and partially South America"* (82).

3.2.3. TRANSPORTATION.

Plain, Train Ships, and Automobiles, The climate impact on transportation. The UK transport, which is the most polluting sector, needs to take some serious decisions. The UK government chose to intervene to prevent the collapse of Flybe (Europe's biggest regional airline) and give the green light for the high-speed rail project, HS2. *"Decarbonizing transport would eliminate 26% of UK CO2 emission.* But Prime Minister Boris Johnson recently said that doing this poses *"difficult and complicated"* questions. On this, Johnson is almost certainly right.

Aviation is one of the fastest-growing fossil fuels consumers," *with airlines contributing about 3.5% of all man-made greenhouse gas emissions"*. This might seem small, but a single *"transatlantic flight from London to New York can grow your carbon footprint by as much as the entire heating budget of the average European"*. At high altitudes, contrails, the white lines we see in the sky, are formed in the wake of aircraft. These high-altitude clouds are too thin to reflect much sunlight, but the ice crystals inside them can trap heat. *"Flights warm the atmosphere by more than the contribution of their CO_2 emissions alone"*. Developing HS2 will mean construction and ongoing maintenance. These all need energy and material investments, manufacture, transport, and use energy. That could increase the carbon footprint of rail by between 1.8 and 2.5 times. For aviation, the same infrastructure requirements are responsible for a 1.2–1.3 increase, with road transport showing a 1.4–1.6 increase. *"This analysis will show the differences between net-zero and absolute net zero emission."*

Comparing life cycles. A life cycle approach gives a better understanding of where emissions are occurring and compares transport modes on a much more level. This helps us understand that most greenhouse gas emissions come from flying and driving, whereas in rail travel, emissions are produced by building the infrastructure itself. Emissions from operating trains are generally lower. But there are still emissions from the manufacture and maintenance of renewable energy technologies to consider. *"Being able to accurately compare the energy requirements and emissions of different transport options is the first step towards addressing their climate impact"*. Governments need to consider and balance the true climate impacts of a transport project (83).

EU's Transport Won't Cut Carbon Far Enough.

Reducing pollution from transport, which accounts for *"almost 25% of European greenhouse-gas emissions"* (globally the percentage of pollution from transportation varies between 22% and 28%), has so far been one of the most difficult tasks for policymakers.

If oil companies were required to buy carbon allowances for the diesel and gasoline they sell to drivers the cost of filling up vehicles would also increase, according to a study by Cambridge Econometrics. *"The European Union's carbon market is the cornerstone of the region's climate policy. EU vehicle CO2 standards and greener taxes at the national level are what's needed to decarbonize road transport."* The European Commission is looking at a carbon-reduction goal of 50-55% by 2030. That would drive up the price of **EU carbon allowances** to nearly 90 euros ($101) a ton compared with the current price of about 27 euros a ton. Making fossil fuels more expensive will be important in the long run (84). The same issues are common for other developed and developing countries. Correlating the reduction of pollutants, CO2, and methane at the global level is the biggest problem of all nations (84).

Electric Vehicles emit fewer emissions and are better for the environment.

A Facebook post claims that electric car batteries weigh approximately

1,000 pounds and require more than 500,000 pounds in raw materials to make (One pound = 0.456kg).

"Driving an electric car won't save the planet," the meme states? Creating one electric battery requires digging up, moving, and processing tons of raw materials, evaluating the batteries' impact on the environment is more complex than the Facebook post suggests. The positive effects of electric cars outweigh any negative impacts from sourcing lithium, to manufacture car batteries. When talking about vehicle emissions, there are two types: *"direct and life cycle"*. **Direct emissions** are emitted through a car's tailpipe and contribute to smog. **Life cycle emissions** are related to fuel or battery manufacture processes and vehicle production, distribution, use, and recycling or disposal. The U.S. Department of Energy says calculating this type of emission is complex but that electric cars produce fewer life cycle emissions than traditional vehicles. Battery production from mining to be installed in the car needs to be driven by about 60,000. miles to offset the carbon footprint used in the production process. For instance, lithium mining in Chile, transported to North Korea, and China to be processed, then shipped to the USA for car installation. How much energy (based on fossil fuel) is required for transportation only? plus extraction and manufacturing processes? Electric vehicle*(EV) production has a long way to go to reach absolute net-zero carbon emission"*. Pollutants from cars that use gas *"directly or indirectly harm human health and the environment"*. Shifting pollutants away from tailpipe emissions to a central location, like a power plant, makes the pollution easier to manage, Lin said. A National Renewable Energy Laboratory report in 2016 found that *"the U.S. transportation sector consumed more than 13 million barrels of petroleum a day in 2014"*. Plug-in electric vehicles (PEVs), can reduce our national dependence on petroleum. Forbes reported that *"lithium-ion batteries are recycle-friendly and can be made into even more batteries. recycling is not only necessary but also urgent."* As electric vehicles become more popular, the battery-recycling market could expand. Research from the National Renewable Energy Laboratory found that *"lithium-ion car batteries may last up to 15 years in moderate climates"* (85).

Tesla's Million-Mile Battery Will Fuel A New Green Energy Boom The million-mile battery for electric vehicles (EVs) could hit the market very soon, giving a boost not only to vehicle ownership but also to renewable energy generation. While the million-mile battery will outlast whatever car it is placed in, it could still be put to good use after its initial purpose. Tesla is reportedly set to launch a million-mile battery as soon as early 2021.

China's battery manufacturer is ready to produce a battery that could last more than 1 million miles. *"As a portion of the carbon footprint is emitted during the production, the utilization of such vehicles exceeds 1,000,000. miles reduces the lifetime carbon footprint per mile traveled. Furthermore, battery recycling has the potential to further* reduce *emissions as components of a battery pack can be captured and reused, displacing much of the need for raw material mining and the associated emissions".* Battery packs can also serve as energy storage for renewable energy generation. For renewable energy, used batteries can be a better choice than building a new energy storage system (86).

Britain to ban new petrol and hybrid cars from 2035.

Prime Minister Boris Johnson is seeking to use the announcement to elevate the United Kingdom's environmental credentials after he sacked the head of a Glasgow U.N. Climate Change Conference planned for November 2021, known as COP26. *"We have to deal with our CO2 emissions, and that is why the UK is calling for us to get to net-zero as soon as possible, to get every country to announce credible targets to get there, that's what we want from Glasgow,"* Johnson said at a launch event for COP26 at London's Science Museum, alongside broadcaster and naturalist David Attenborough. *"We know as a country, as a society, as a planet, as a species, we must now act."*

The mayors of Paris, Madrid, Mexico City, and Athens plan to ban diesel vehicles from city centers by 2025. France is preparing to ban the sale of fossil fuel-powered cars by 2040 and Norway's parliament has set a non-binding goal that by 2025 all cars should be zero emissions.

The UK government said it was providing an extra 2.5 million pounds ($3.25 million) to fund the *"installation of more than 1,000 new charge points for electric vehicles"* on residential streets. It has also provided investment for the development of electric vehicle technology.

Electric cars accounted for a 44.3% share of Norway's new car sales in January 2020, rising year on year (87).

Trucks And Large Commercial Vehicles.

In the USA a coalition of states is following California's lead in setting goals *"to jump-start a transition to electric-powered trucks, vans, and buses"* to reduce greenhouse gas emissions. The 15 states, plus Washington D.C announced in July 2020, that they've agreed to have 100% of all new medium- and heavy-duty vehicles sold be zero-emission by 2050, and of 30% zero-emission vehicle sales by 2030. California's Air Resources Board announced in June 2020, requiring that all new commercial trucks and vans purchased must be zero-emission by 2045. with milestones along the way. Or the states could focus more on subsidies and incentives, as well as investment in charging infrastructure. A few examples of companies to produce electric trucks. **Paccar Inc.** takes a long view of hydrogen fuel cells, the company is more of a leader in both battery and fuel cell electric powertrains. *"To date, the company has deployed over 60 battery-electric, hybrid, and hydrogen-powered trucks.*

Kenworth recently completed a project with **Toyota Motor Corp.** building 10 heavy-duty Kenworth T680s equipped with twin fuel cell stacks designed for Toyota Mirai passenger cars. Some are in use with customers in the Port of Los Angeles. *Today hydrogen is $12 to $13 per kilogram,"* for it to be efficient, it needs to be in the $2- to $3-per-kilogram range. *"By contrast, startup **Nikola** Corp.* plans Class 8 fuel cell truck production in 2023 at a new plant in Coolidge, Arizona. Nikola is also working on a network of hydrogen fueling stations as part of its plan to offer trucks, hydrogen fuel, and maintenance. Battery-electric trucks for sale in 2021. Also, on the battery-electric side, Kenworth and Peterbilt will build medium-duties for sale or lease in 2021.

Navistar International Corp. is working with autonomous startup TuSimple to launch a Class 8 truck in 2024. *"Without fanfare, Kenworth showed a Level 4 autonomous truck at the 2020 Consumer Electronics Show in January."* This is the trend for vehicles and trucks free of emission on pipe tables. The question is: Is it too late to stop the Climate Crisis? (88).

3.2.4. AVIATION.

Hydrogen fuel could revolutionize airlines.

Global airline travel has grown over the decades, and with it, so have the industry's carbon emissions. Can the airline industry shrink its carbon footprint? which currently makes up 3% of all U.S. GHG emissions and 3.5% globally. The answer hinges on the development of alternative fuels. In September 2020, European aircraft maker Airbus *announced it would evaluate three concept planes, powered by hydrogen."* The goal is that airplanes could enter commercial service by 2035. Hydrogen's incredibly energy-dense, much more than current battery technology. It's also plentiful and burns cleanly, it is a good option, producing no carbon dioxide or carbon monoxide. Fission energy to produce hydrogen can be an appropriate electric source. There are issues with hydrogen, which need to be solved. Universal Hydrogen is developing a type of *"capsule technology that would enable either liquefied or gaseous hydrogen to be shipped".* The start-up plans to begin service with regional airlines in 2024. Airbus is also looking at hybrid hydrogen-electric planes and all-electric planes. *"Airlines aren't necessarily motivated by the environment,"* said Ryerson of the University of Pennsylvania. *"If making a profit and reducing their greenhouse gas emissions works together, they will invest in reducing greenhouse gas emissions. But if those two things are not tied, they will favor making a profit." Green flying from now on will be a luxury, not many will afford it"*(88).

Zeroavia Startup. ZeroAvia has completed what it calls the world's first hydrogen fuel cell-powered passenger aircraft flight, in a six-seater commercial Piper aircraft. The flight was a part of the Zero Avios UK government-backed HyFlyer project (89). **Ammonia Can Work Like Jet Fuel**.

A government-startup collaboration in the U.K. finds ammonia could power jets.

The new design is based on using carbon-neutral heat to turn ammonia and hydrogen into jet fuel. The project is a collaboration between startup Reaction Engines and the U.K. Research and Innovation's Science and Technology Facilities Council (STFC). Ammonia is one of the most plentiful chemicals that humans produce because it's massively used as a fertilizer, but requires considerable energy to make. *"The density of liquid ammonia allows for conventional aircraft configurations to be used, resulting in a zero-carbon jet that could start serving the short-haul market."* This represents one more solution for green aviation (90).

Scientists Turn CO2 Into Jet Fuel

"Oxford University scientists have successfully turned CO2 into jet fuel". This means aircraft run with net-zero emissions. *"The team heated a mix of citric acid, hydrogen, and an iron-manganese-potassium catalyst to turn CO2 into a liquid fuel capable of powering jet aircraft."* The process requires much more widespread use of carbon capture. For net zero emissions, the capture and conversion systems would have to run on clean energy.

It might also be one of the most viable options for fleets. This conversion process would let airlines keep their existing aircraft and go carbon neutral until they're truly ready for eco-friendly propulsion (90a).

3.2.5. SHIPPING

U.N. shipping agency, the International Maritime Organization (IMO), intends to reduce GHG emissions by 50% from 2008 levels by 2050. At least $1 to 1.4 trillion of investment is needed to meet U.N. targets. *"The global shipping fleet, which accounts for 3 % of the world's CO2 emission. "About 90% of world trade is transported by sea"*, using over 50,000 ships. Studies show that industry needs an average of $50 billion to $70 billion annually for 20 years. *"Our analysis suggests we will see a disruptive and rapid change to align to a new zero-carbon system, with fossil fuel aligned assets becoming obsolete or needing significant modification,"* said Tristan Smith, a reader at University College London's (UCL) who was involved in the study (91).

Many European banks exit from providing finance for the shipping industry.", leaving a capital shortfall of tens of billions of dollars annually. Around 87% of investments needed would be in land-based infrastructure and production facilities and the remaining 13% of investments are related to the ships themselves. The shipping industry is responsible for an immense amount of pollution. *"That's partly because many vessels burn a heavy form of* oil called bunker fuel that's rich in sulfur." *Also,* it remains a major source of nitrogen oxides, another powerful greenhouse gas. *The goal is for shipping to reduce its carbon-dioxide intensity by at least 40% by 2030 compared to 2008 levels".* Developers are testing the use of hydrogen to power ships. The UN says the *"first net-zero ships must enter the global fleet by* **2030.***"* Swiss co. *"ABB sees short-distance shipping as the first adopters of the fuel cell technology,"* said Juha Koskela, division president, ABB Marine & Ports. *"Green hydrogen fuel costs around 4-8 times the price of very low sulfur fuel oil",* estimates by risk management firm DNV GL find.

In Norway, *"cruise ships and ferries sailing through the country's heritage-protected fjords must be emissions-free by 2026."* The company is also working on a separate hydrogen project for wind installation turbine vessels. Britain's Felixstowe port is looking into hydrogen, based on offshore wind farms and a nuclear power plant (92).

Cargill Looks to Use the Wind to Cut Carbon in Shipping

Cargill, one of the world's biggest charterers of ships, is working with technology partners to fit sails on vessels in its fleet to cut carbon emissions through harnessing wind power, the U.S. agribusiness Cargill said on 28 Oct. 2020. BAR Technologies and naval architect Deltamarin to develop *"wing sails that reach up to 45 meters in height",* and *reduce CO2 emissions by up to 30%". This technology is helpful in our journey to zero-carbon vessels,"* The project is in the design phase with the first vessels expected on the water in 2022 (93).

3.2.6. ALUMINIUM, STEEL, AND OTHER PRODUCTS.

Other big GHG emitters deserve to be mentioned, due to the industry's fight to comply with neutral emission by 2050.

Aluminum producers' race to go green may fracture the market**.** Tesla is ordering giant aluminum casting presses for its assembly line in Germany. Aluminum is one of the materials of choice for the electric vehicle because of its *"lightweight and strength".* Indeed, aluminum is going to be one of the metals critical to the unfolding green revolution, for its use in renewable energy sources, particularly solar panels, and battery storage.

"The sector accounts for almost 3% of global emissions." But this widening rift between "green" (green energy), and "black" (energy based on fossil fuels) aluminum risks fracturing the markets. *"Aluminium smelting is an"* **energy-intensive business.** The average comes in at *"around 10 tonnes of carbon per one tonne of aluminum produced"* but the global range can be anything from 4 to 18 tonnes.

Elysis, a joint venture between Rio Tinto and Alcoa, is already there, using a process that eliminates all direct greenhouse gas emissions from smelting. However, many producers, particularly those in China, are going to find the road to zero emissions a long and winding one, if not impossible. The emerging split between *"green"*, and *"black"* aluminum is only going to widen. The same issue is with **steel** products around the producers and industries in developed countries. *"Russia, China, and Brazil are dominating the steel mining and production process."* Emitting rates for steel production vary with the country and are between 5% to 10%. *"Globally the rate fluctuates around 8% to 10%".* Steel global production is heavy energy consumption, going neutral by 2050 remains in doubt (94). Readers may not believe it, but **fashion** industries are among big emitters too, with emitting rates between 2.5% and 3% of global emission of GHG. Also, I have to emphasize petrochemical products (like plastic, detergents), concrete and wood manufacturing, food processing plants, etc., as constant emitters of GHG. (4).

3.2.7. DEFORESTATION, PEATLAND, AND PERMAFROST

Deforestation, General Notes.

Tropical rainforests are disappearing at an alarming rate, and according to a new report by the non-governmental organization Rainforest Foundation Norway (RFN), humans are to blame. *"The world's dependence on coal, farming, soy, and palm oil. Mining has resulted in two-thirds of Earth's tropical rainforests being destroyed."* The remaining rainforest being put *"closer to a tipping point,"* Tropical rainforests once *"covered 14.5 million square kilometers (msqkm), 13%, of Earth's surface"*, according to RFN, but now, just 33% of that remains intact. Of the original area tropical rainforests once occupied, *"34% is completely gone, and 30% is suffering from degradation"*. All that remains is roughly 9.5 msqkm, *"and 45% of that is in a degraded state,"* the study says. The findings are more than alarming.

"Humans are destroying these forests, cutting their ability to store carbon, cool the planet, produce rain and provide habitats". The world depends on tropical rainforests to provide these services. The 17 years study from 2002 up to 2109 concluded: *"that since 2002, the area of rainforest lost is greater than the size of France"*. To blame are "human rampant economic development." Agriculture, energy consumption, trade, and the production of soy, palm oil, cattle, logging, and mining have been the largest threats over the past century. Tropical rainforests constitute more than half of the Earth's biodiversity and store more carbon in living organisms than any other ecosystem, according to RFN. Of the world's remaining tropical rainforest cover, 70% is found in the Amazon and Africa rainforest. Congo holds more than half of Africa's tropical rainforests. Tropical rainforests put *"the world at risk of future pandemics. Deforestation is disturbing nature's natural virus protection mechanisms, exposing humans at risk of pathogens spreading."* The United Nations issued a joint statement to recognize *"the right to a healthy environment. We are faced with a triple environmental crisis: climate change, loss of biodiversity and pollution"*. The saving of this ecosystem depends on our and future generation's rights (95).

"Deforestation in the Amazon destroyed an area bigger than Spain from 2000 to 2018",* wiping out *"8% of the world's biggest rainforest"*, according to a study released Tuesday, Dec. 7, 2020. Since the turn of the millennium, *"513,000 (sqkm) (198,000 sqm)"* of the rainforest has been lost. *"In 2018 alone, 31,269 sqkm of forest were lost across the Amazon, the worst annual deforestation since 2003."* The Amazon stretches across eight South American countries: Brazil, Colombia, Peru, Bolivia, Ecuador, Venezuela, Suriname, and Guyana, and the territory of French Guiana. Brazil, which holds about 62% of the Amazon. 425,051 sqkm from 2000 to 2018, which represents 82.85% of the total area lost in the Brazilian Amazon. In one year, 2019-2020, Brazil lost *11,088 sqkm.* *"Land grabbers and illegal loggers and miners"* are the main drivers of deforestation on indigenous reservations, where the rainforest has been protected by law. Environmentalists *"who blame President Jair Bolsonaro for advocating the development of the Amazon."* Since 2017 the deforestation area has increased from year to year to year. On top of the fire destroyed an additional considerable area. Brazil risks the pandemic public health crisis and causes lasting harm to the forest and indigenous communities. *These are areas in the Amazon where 60% of all deforestation occurs. "But Jair Bolsonaro's government appears set against the aims of the prosecutors."*

Investment funds managing close to $4 trillion in assets called on Brazil, June 23, 2020 *"to halt deforestation of the Amazon". Europe, Asia, and South America"* said that the government in Brasilia was using the COVID-19 crisis to *"jeopardize the survival of the Amazon.* Environmentalists warn 2020 is the most destructive year ever for the world's biggest rainforest. **Joe Biden,** the newly elected President of the USA has raised the issue of tropical deforestation as a *"critical climate issue". "$20 billion in international funding to save the Amazon in Brazil."* In doing so, the scale and impact of ending tropical deforestation in the Amazon and elsewhere are immense. *"There is no solution to the climate and biodiversity crises without ending deforestation".* One way is to introduce *"worldwide regulatory measures to protest Amazon forests as well as other tropical forests in Africa and Southeast Asia" (96).*

Southeast Asia's Rainforests

In *"Malaysia and Indonesia"*, the largest tropical forests in Southeast Asia, during health crises *"the lockdown is impacting forest protection, and companies may take advantage of the lockdown to expand and clear the forest,"* The world lost 12 million hectares (30 million acres) of tropical tree cover in 2018. From 2001-2018, Malaysia lost rainforests equal to 7.7 million hectares. In recent years, palm plantations in Malaysia and Indonesia, which have *"the world's third-largest tropical forests",* have come under scrutiny due to logging, land clearing, fires, and labor abuses, RFN said. (RFN).

Central Africa

In Central Africa, the world's second-largest tropical rainforest, the protection from deforestation is part of an experiment in the Congo

Basin: giving power to the people in an attempt to preserve the rainforest. Deforestation rates have accelerated, the Congo Basin step in the fate of the Amazon rainforest. Since 2010, nearly 11 million acres, (4.58 million hectares) of *"primary forest", the oldest, densest, and most ecologically"* significant kind, have been lost in the Congo Basin to *"logging, agriculture, mining, and oil drilling"*. Experts say *"that if nothing is done, the more than 400 million remaining acres will disappear by the end of the century."* There is time to head off disaster. The Democratic Republic of Congo owns nearly 60% of the forest. In early 2016, the government passed a law giving to individual villages" *185 million acres of forest, with the expectation that local ownership leads to sustainable management* (97).

Other areas of the planet encountered uncontrolled deforestation in South America and Europe. For instance, in Romania, the virgin Carpathian forests are cut in with no control from the Government or Government Institutions. All the forest wood takes the road outside the country for the western state's industries of Europe, crossing the border without any difficulties. That all tells to every one of the grafts practiced in Romania from the moment the trees are cut until it arrives in the West. Even the European Commission has expressed concern about the high rate of deforestation in Romania and asked the Government to take action. For 31 years since the fall of communism, no leftist or rightist Government has done anything to stop the massive thief of Romanian lumbers.

Peatland and Permafrost.

"Peatlands cover just about 3% of the global land area" but *"they store almost 25% of all soil carbon and so play a huge role in regulating the climate"*. We have the most accurate map yet of the world's peatlands, their depth, and how much GHG they have stored. *Global warming will soon mean that these peatlands start emitting more carbon than they store"*. The peatlands developed in northern tundra and taiga areas where they have helped cool the global climate for more than 10,000 years. Now, large areas of *"perennially frozen (permafrost)"* peatlands are thawing, releasing the freeze-locked carbon back into the atmosphere as *"carbon dioxide and methane"*. The study authors put together a map, which we can use to estimate how the peatlands will respond to global warming. Peatlands

cover approximately 3.7 million square kilometers." *If it were a country, "Peatland" would be slightly larger than India"*. These peatlands also store approximately 415Gigatons (billion tons) of carbon, as much as is stored in all the world's forests and trees together, *or 11 times more than the human activities released in 2019 (38 Gigatons)."* That is huge, can and may not be ignored. Permafrost peatlands are in Western Siberia and around Hudson Bay in Canada. *"These unique environments and ecosystems will be fundamentally changed"* as the permafrost thaws. These changes will cause more CO_2 and methane to be released into the atmosphere, and will also lead to large losses of peat into rivers and streams, *"which will influence both the food chains and biochemistry of inland waters and the Arctic Ocean. The only way to stop permafrost thaw is to limit global warming."*

Peatlands are found in an estimated 180 countries. Many of them have not been recognized and are not yet properly mapped. Levi Westerveld/GRID-Arendal, CC BY-N.

Peatlands exist on every continent, even in Antarctica. In the U.S. they are found in many states, including Maine, Pennsylvania, Washington, and Wisconsin. *"Humans have used peat for centuries as a fuel, and also to flavor whiskey." Temperatures soared 10C above average across much of permafrost-laden Siberia in June 2020"*. In many regions of the Arctic, rapid permafrost thawing promotes microbial activity that releases GHG. Peatlands exist on every continent, even in Antarctica. In the U.S. they are found in many states, including Maine, Pennsylvania, Washington, and Wisconsin. *"Humans have used peat for centuries as a fuel, and also to flavor whiskey." Temperatures soared 10C above average across much of permafrost-laden Siberia in June 2020"*. In many regions of the Arctic, rapid permafrost thawing promotes microbial activity that releases GHG.

Recent *"wildfires like those in Russia are known to release as much carbon in a few months as total human carbon dioxide emissions in an entire year"*. This is more than scary for global warming. Human activities are also increasing GHG releases. In the United Kingdom, for example, extracting peat for use in gardening has caused peatlands to emit an estimated *"16 million tons of carbon every year, the same as 12 million cars"*, above 3°C-4°C more than that of the Paris agreement (98).

Most peatlands in Indonesia have been destroyed to build palm oil plantations. *"Indonesia and Malaysia, peatland draining results in total annual emissions equal to those of nearly 70 coal plants. Peatland degradation due to human activity accounts for 5-10% of annual carbon dioxide emissions"*. Peatlands are not included in most earth system models" used for climate change projections. *The study* *"highlights the need to integrate peatlands into these models"*. Net changes in peat carbon over the *"next 80 years"* could range from a *"gain of 103 billion tons to a loss of 360 billion tons"*. Peatlands should also be considered in *"integrated assessment models"* that researchers use to understand climate change impacts. The first step is to raise awareness around the world of this precious natural resource (99).

3.3. CARBON DIOXIDE (CO2)

Level Past Pandemic

According to the US agency (NOAA), the monthly average CO2 concentrations, recorded at the Mauna Loa Observatory in Hawaii, were 417.1 parts per million (ppm) 2020 year compared to 413.33 ppm in April 2019. **It's the highest concentration since records began in 1958.** A decade ago there was 393.18 ppm, NOAA reported, an increase of 23.92 ppm. *"During the 1960s, the increase over one year was an average of 0.9 ppm, which has risen to an average 2.4 ppm a year in the past decade"*. Coronavirus pandemic could lead to a drop in emissions but that without structural change, *"have little impact on the build-up of CO2."* Emissions would have to *"start falling by an average of 7.6% per year to achieve the 1.5C temperature goal"* (100).

Carbon Credits. Helping to Cut Emission?

Most scientists are, however, under no illusion that the *"climate crisis is one of mankind's gravest existential threats"*. Climate change could *"cost the global economy at least $1 trillion over the next five years in crumbling infrastructure, reduced crop yields, health problems, and lost labor as per the Carbon Disclosure Project (CDP)"*.

"Large corporations know this and have been falling over themselves

pledging to achieve net-zero or even carbon-negative status within our lifetimes". Their intentions might be good, but their oft-preferred route for achieving this goal, purchasing carbon offsets, is likely to fall far short of the ideal, per United Nations Environment Program (UNEP). Fossil fuel companies, some of the biggest emitters of carbon, are conspicuous by their *"absence on this list of more than 100 companies"* that have committed to lowering their carbon footprint. But merely *"buying carbon credits in exchange for a clean conscience"* while these companies carry on *"powering their operations using high-carbon fuels"* will no longer suffice.

Scientists, activists, and concerned citizens are now questioning how companies are using *"carbon offsets as a free pass for inaction."* The world needs to lower annual emissions by *"29-32Gigatonnes" of equivalent carbon dioxide (CO2e) by 2030. "That's ~5x (approx .5 times) the current commitments by companies, organizations, and governments".* We need to lower our GHG emissions *"by 45% over the next decade if we are to avert catastrophic planetary changes."* Perhaps *"renewable energy credits, or RECs",* are a better alternative. Whereas a *"carbon offset represents an action"* that effectively sequesters carbon, *"RECs are like a property deed."* The only foolproof way of directly cutting GHG emissions is to go the Amazon or Google way. Amazon has revealed plans to invest in four new renewable energy projects, and meanwhile, Google has signed up to $2bn wind and solar investments.

Carbon Footprints

"Big Four accounting firm KPMG is positioning a new blockchain-based accounting capability to help companies in their efforts to reduce their environmental footprint."

On October 6, 2020, KPMG announced a patent-pending blockchain-based capability, the *"Climate Accounting Infrastructure (CAI)".* The initiative aims *"to help organizations better measure, mitigate, report, and offset climate-changing emissions",* according to a KPMG press release. CAI will track an organization's emissions and offset records on a blockchain, which will be integrated into an organization's existing systems. Real-time environmental data and advanced analytics will be recorded on CAI as well to model the impact of climate risks on financial performance, according to the firm. The rationale behind the project is that capital market investors and consumers increasingly expect *"organizations to meet environmental, social, and corporate governance demands (ESG)"*. To meet those expectations, KPMG is positioning the technology to track sustainability and comply with climate change legislation. *"Trusted reporting capabilities, such as those enabled by CAI will be critical to meet stakeholder expectations and to comply with emerging regulations,"* said Arun Ghosh, KPMG's U.S. Blockchain leader, in a statement. In other words, all financial operations of a company will be processed by an algorithm to verify if the operation meets ESG requirements (101).

UK Invisible Footprints

UK efforts to reduce GHG emissions are being seriously undermined by the failure to ***"account for rising carbon*** *output in the countries where our goods are manufactured"*. Almost 50% of the UK's carbon footprint, (the total amount of GHG released in the production and consumption of all the goods and services), now comes from emissions released overseas for *"imported products."* The same issue is countable for any country in the world. These "invisible" overseas emissions are *"not currently included in the UK's 2050 net-zero targets"*, while international aviation and shipping emissions are also excluded. Between 1990 and 2016 total emissions inside the UK's borders had been reduced by 41%, but the consumption-based footprint only dropped 15%. This is due to goods and services coming from abroad. *"Overall, 46% of the UK's carbon footprint in 2016 came from emissions released overseas."* This *"highlights the importance of addressing carbon-intensive imports such as animal feed and fossil fuels.*

Dr. Stephen Cornelius, the chief climate change adviser at WWF, said: *"Climate change is a global problem that needs a global solution"*. As an influential nation that has shown it can act as a global leader on climate change before, *"we can take responsibility for emissions that are down to UK demand alone"*. This is a typical country issue with carbon emission, the problem is that this issue applies to all countries, being a global issue (102).

Footprints and Rich. To control the climate crisis we need to reduce GHG emissions equivalent to 2.5 tonnes of CO_2 per person per year... A study shows *that only about 5% of EU households live within these limits"*. *What about the other 95%?* The vast majority of EU citizens are using far more than their fair share. In the EU, the average carbon footprint is 8 tonnes of CO_2 per person, which must fall to 33% over the next decade. *"Households in the top 1% in the EU have carbon footprints that are 22 times larger than the limit of 2.5 tonnes."* On average, these people emit GH equivalent to 55 tonnes of CO_2 per person per year. In most developed countries the situation is the same. So who are these top emitters? We know they're relatively wealthy with an annual net income of around €40,000 per person. The top 10% in the EU account for 27% of the total EU carbon footprint, contributing more than the bottom 50%. *"For 95% of EU households"* requires radical change. (103).

Carbon Price. *"Price per metric ton of carbon emissions hits $20"* and must go up. That is according to the IHS Markit Global Carbon Index. The index tracks the most widely traded carbon credit futures contracts in the U.S. and Europe. *"10 U.S. states and the European Union"*, have effectively put a price on carbon emissions. Some have required companies and utilities to purchase credits for each ton they emit, which can then be sold or traded to other companies. That is what the price of the Global Carbon Index is about. Some estimates say that *"carbon needs to reach $100 a ton to achieve the emissions-reductions goals of the Paris Agreement."* That is about five times the current levels.

US Carbon Pricing. Morgan Stanley, JPMorgan Chase & Co., and BP Plc. support introduction of carbon pricing in the USA. Using it in Europe has seen some success. The U.S. has similar carbon markets, led by California's, not comparable to Europe's scale.

"Carbon dioxide emissions inflict a cost on society that is not reflected in market energy prices, and a carbon price is the most efficient way to fix it". According to Jesse Keenan, a Tulane University professor "t*his isn't just about Wall Street,"* he said. *"This is about the entirety of the U.S. financial system, and that includes our housing and labor markets as well."* A group of CEOs from some of the biggest U.S. companies, announced in mid-September 2020, it supports market-based carbon pricing to fight the climate crisis (104).

Political Approach. *"Carbon pricing is such a hot potato that politicians are avoiding talking about it"*. In the mid-Oct. 2020, the obscure, Republican-controlled federal agency that oversees the electric grid said: *"it would support the price of carbon in the power sector"*. Translation: *"There's a good chance the US could get a carbon price"*. The Federal Energy Regulatory Commission (FERC) its usual business is to assure that interstate electric grids are operating properly. But in September 2020, FERC's *"commissioners agreed to weigh in on carbon pricing, in response to a proposal from New York grid starting imposing a carbon price on fossil-fired power plants."* The first-of-its-kind proposal will force grid operators to prioritize green energy from wind, solar, nuclear, or other zero-carbon sources, per laws and policies. FERC would need to sign off and become a central player in climate policy. This will be less politically toxic and has a greater chance of survival than a traditional carbon tax. *It's much more subtle"* (105).

The USA and the EU's Carbon Border Tax.

The EU will *"enact a new tax on products from countries that aren't working to reduce their emissions"*. U.S. companies face serious, and costly, disadvantages as they compete for trade in the EU. *"No sense to reduce carbon emissions inside Europe, to then import them from outside."* France began a low-key campaign last year 2020, to push the EU to implement a border carbon tax. The new EU president included a border carbon tax to eliminate the bloc's carbon footprint and provide more than *"$1 trillion in green finance"*. *"The potential implications of such a measure are just beginning to ripple across the globe. Officials in China, the world's largest emitter, worry that it might dampen the country's economic growth. "Several groups in the USA are pushing plans to **implement its carbon tax,** combined with a border tax"*. *"This creates an incentive for other countries. I want to get inside that club,"* says former Federal Reserve Chair Janet Yellen, treasury secretary now (106).

Carbon Levy on Marine Fuels.

The International Maritime Organisation (IMO) wants to halve the sector's greenhouse gas output by 2050 against 2008 levels. *"If unchecked, global emissions from shipping could go up to 130% by 2050, compared to 2008 levels.* The revenue raised by the levy would be used to subsidize low and zero-carbon fuels."Trafigura, which is responsible for more than 4,000 ship voyages per year, acknowledged that a carbon price will have an immediate effect on shipping costs that companies would bear. *"Charterers have to change behavior to reduce emissions, charter more efficient ships and switch to lower carbon fuels" (108).*

Companies to Pay off Carbon Debt.

A new environmental target is connecting *"to paying reparations to victims of past injustice."* We *"eliminate not only current pollution but also to account for climate damage from the corporate past."* The first company to pledge was Microsoft Corp. In January 2020, it announced plans to remove enough GHG to zero out its emissions *"dating back to its founding in 1975, some 27.3 million tons of emission. Now Velux A/S, a Danish maker of roof windows, pledged to eliminate carbon since its founding in 1941.* The idea of measuring historical CO_2 gets very attractive.

For Velux, these payments for emissions through preventing deforestation. Microsoft also backs forest projects, and software giant *"created a $1 billion climate investment fund"* to bolster carbon-removal technologies. Climate Accountability Institute tracks the historical emissions of the world's Big Oil *"The group's data, which goes back to 1965, ranks Saudi Aramco at No. 1 with past emissions equal to nearly 60 billion tons of carbon dioxide (60 Gigatons). ExxonMobil, Chevron, BP, and Shell all rank in the top seven".* The list of companies following Microsoft is increasing (107).

No on Paris Agreement, No on Carbon Exported.

This September 2020, the French government stepped in to block a $7 billion, 20-year agreement that would have sent millions of tons of liquefied natural gas (LNG) from the US to Europe. *"As far as international struggles go, it remains a decorous disagreement between allies".* Yet it highlights the fault lines emerging as the US formally leaves the Paris climate pact on Nov. 4, 2020. The new Washington Biden administration will review all these policies.

"Carbon taxes, or border adjustments, are the next frontier of global trade (106).

3.4. POLLUTION

The Earth is Not a Math Problem.

"The earth, our home," Pope Francis wrote, *"is beginning to look more and more like an immense pile of filth."* That was five years ago IN 2015. Many observers, not only in secular media but *"in conservative"* Catholic circles as well, misunderstood the point of his encyclical on the subject of what he calls *"human ecology, same as climate change."*

"It has (human ecology) distinct but overlapping political, social, economic, and, of course, spiritual and ecological dimensions". Its urgency is such that the Holy Father has addressed many of his recent pronouncements but for all persons of goodwill throughout the world. For Francis, t*he environment* is so valuable, remindings us that it is not an abstract problem. It will not be solved by a *"technocratic mindset"* or schemes like the *"Green New Deal. The measurability of waste is beside the point because the earth is not a math problem".*

If scientists could demonstrate that plastic trash in the ocean helps birds live longer, our throwaway culture would still be disordered.

This is true that we discard without thought, *"the consequences of disposal will be with us for thousands of years".* We read that even the summit of *"Mount Everest is becoming a trash heap"*, and its snow is soaked with microplastics and cast-off water bottles and fragments of synthetic fiber. *"Can you graph the Himalayas?"* The prospect of *human extinction* raised by climate scientists is a fantasy that distracts us from the real question: *"What sorts of human beings will live in an utterly spoiled world, centuries or millennia after their ancestors traded beautiful mountains for ones made of Paw Patrol action figures?"* (109).

Pollutants. Air pollution causes some *"7 million premature deaths each year"*, of which 600,000 are children under 15 years old. The most harmful pollutants are **microscopic** particles in the air. These are produced by farming, industrial activity, wood burners, and road transport. The very finest particles, released by fuels, are the most dangerous. This particle penetrates deeper into the respiratory system and from there into the bloodstream. **Ozone** is also a pollutant. In the stratosphere, it exists naturally, at ground level, it's a health hazard. It is created by reactions between nitrogen oxides, (NO) and volatile organic compounds in the presence of sunlight. These compounds come especially from exhaust fumes. Ozone can damage vegetation and cause stress for people with asthma and respiratory illness. Another gas, **nitrogen dioxide** (NO2), forms quickly in high-temperature combustion, in-vehicle engines, and power plants. It reacts with vapor to create nitric acid, which can worsen bronchitis. Pollutants such as sulfur dioxide, benzene, ammonia, and toxic metals like lead and cadmium can also affect the lungs, eyes, digestive tract, or nervous system.

Note: *'Air pollution is back to or exceeding pre-pandemic levels'* in 80% of the UK despite continued lockdowns, according to a new study. (on Dec. 9, 2020).

One of the causes is *"a surge in private car use"* as people avoided public transport during and after lockdown. Toxic air has accelerated the deaths of thousands of Covid-19 victims, and even after the pandemic ends (Wikipedia).

Air Pollution Costs Europe Cities $190bn a Year: (analysis).

Air pollution costs inhabitants of European cities more than 160 billion euros ($190 billion) each year due to long- and short-term health impacts said in research published Oct. 16, 2020. Pollution from fossil fuels in *"2018 cost the average citizen 1,250 euros about 4% of their annual income"*.

Pollution in Paris-France.

Reducing air pollution in European cities should be a top priority. The present Covid-19 pandemic has only underscored this." The study looked at more than a dozen health factors linked to air pollution in cities to quantify the "social cost" of exhaust fumes and factory smoke on populations. The authors put the cost for *"130 million citizens in cities studied at 166 billion euros"* in 2018. **London** had the highest social cost from pollution totaling 11.38 billion euros lost welfare. **Bucharest** (6.35 billion euros lost) and **Berlin** (5.24 billion euros lost). In July 2020, the Air Quality Life Index showed that air pollution *"cuts life expectancy for every man, woman, and child on Earth by nearly two years. Nearly 25% of the global population lives in just four south Asian countries that are among the most polluted -- Bangladesh, India, Nepal, and Pakistan. These populations would see their lifespan cut by five years on average"* (110).

China Says Environment is still grim despite five years of progress.

China's environmental conditions are "grim", falling short of public expectations. There was still a long way to go, said Zhao Yingmin, the vice-minister of ecology and environment. China had met a series of targets on smog, water quality, and carbon emissions in the last five years from 2016. China remains dependent on heavy industry and coal. **10,000 deaths** from air pollution in China have been prevented by coronavirus lockdown, study finds. A study from the University of Birmingham, published in *"The Conversation",* revealed a 63% reduction in *"nitrogen dioxide concentrations"* in Wuhan, the country's epicenter of COVID-19, while residents were ordered to stay at home *between 23 January and 8 April. Without Covid per year, it will be* **48,666.** *death due to pollution.* **This is a number to be remembered"** (111).

China Must Commit to EV battery recycling. China needs to step up the recycling and repurposing of batteries for electric vehicles to ease supply strains and **curb pollution and carbon emissions**. The manufacturing of batteries is energy- and carbon-intensive.

Pollution in China.

"How the government responds will have huge ramifications for Xi

Jinping's 2060 carbon neutral commitment," Greenpeace said *12.85 million tonnes of EV lithium-ion batteries will go offline worldwide between 2021 and 2030, while more than 10 million tonnes of lithium, cobalt, nickel, and manganese will be mined for new batteries.* Repurposed batteries could be used as backup power systems, and would save 63 million tonnes of carbon emissions. **China, the world's biggest EV user, and EV battery manufacturer** have launched its battery recycling schemes.

India's capital chokes on 'severe' smog as farm fires soar. Delhi turns into a toxic soup. New Delhi was blanketed in a noxious haze as air pollution levels soared to *"severe"* levels. The air quality index at Delhi's 36 pollution monitoring sites was between 282 and 446, pushing levels into the "severe" category, the Central Pollution Control Board said. The *"good"* category is between 0-50. Scientists warned in 2020, the pollution season would make *"Delhi's 20 million residents"* more vulnerable to the coronavirus (112). **Latest:** New Delhi, the capital city with the worst air quality worldwide, suffered its most toxic day in a year on November 5, 2020, recording the concentration of poisonous PM2.5 particles at 14 times over the World Health Organisation safe limit. *"At this time in Delhi, coronavirus and pollution are causing major havoc.* The overall air quality index (AQI), which includes other pollutants besides PM2.5 particles, *"crossed 460 on a scale of 500, the worst since Nov. 14, 2019."*

Residents in Lagos (Nigeria) are paying for the city's worsening air pollution with their lives. The city which is home to **over 20 million people** is infamous for congestion and snaking lines of traffic jams. Data from a **World Bank study** estimates *'air pollution resulted in 11,200 premature deaths in 2018' alone.* Children under five, accounting for 60% of deaths in 2018.

In monetary terms, the cost was up to *"$2.1 billion or 2.1% of the state's GDP."* The matter of particle concentration in Lagos is *"nearly seven times higher than the WHO."* Levels of *sulfur were 150 times higher than the permitted European limit (113).*

California's Air Quality is worse than India's. That's not good in a pandemic.

As of 2:30 p.m. PDT Monday, Aug. 24, 2020, about one-third of the

Golden State was deemed to have air unhealthy for all members of the general public, according to the Environmental Protection Agency's Air Quality Index. That assessment included highly populated spots like San Francisco, Bay Area... *"At times during the past few days, parts of the Bay Area have been blanketed by a thin layer of ash while enduring the worst air quality in the world. Same Monday afternoon's rating of **548 particles**, west of San Jose was three times higher than the closest global figure". The concentration of the tiny particles (PM2.5) in the Bay Area is roughly five times the daily average limit set by the EPA. It's worse in the Bay Area now than megacities like New Delhi, which are known for poor air quality."* Even healthy people are reporting headaches, bloody noses, etc. The EPA calculates a daily Air Quality Index based on five major pollutants regulated by the Clean Air Act: ground-level ozone, particle pollution (also known as particulate matter), carbon monoxide, sulfur dioxide, and nitrogen dioxide. The Air Resources Board advises people in the affected areas to stay inside with their windows and doors shut, run air conditioners *"The prevalence of smoke only makes residents more vulnerable to the highly contagious disease."*

"Four west coast cities in the US currently rank in the top 10 for worst air quality in the world", as wildfires rage up and down the western seaboard, cloaking the entire region in smoke.

Portland in Oregon, and Seattle in Washington, hold the No 1 and No 2 spots, while San Francisco and Los Angeles sit at four and six. Collectively, with the smoke from the wildfires, these four cities have knocked every city in China out of the top 10 for worst air quality.

Residents in Oregon And Washington are now experiencing what their California counterparts have gone through *"since mid-August until mid-Sept. 2020"*. Air quality in five major cities in Oregon was the worst on record. The resulting blanket of ash and smoke has made the *"region's air quality among the worst in the world"*. The hardest hit is Oregon, where tiny bits of smoke and ash known as particulates have reached the highest levels on record *"Air in Oregon was rated hazardous"* according to air quality standards, *"and in Bend, the air quality index topped **500**, exceeding the air quality scale altogether"*. Satellite images show that some of the smoke from the fires, traveling on the jet stream at high altitudes, has wafted east as far as New York and Washington, D.C., according to the National Weather Service (114).

Entire West Coast With the Worst Air Quality on Earth

Research has linked particulate matter from wildfires to heart and lung problems, increased hospital visits. "*The EPA recommends residents stay indoors with filtered air, keep physical activity levels low, and wear an N95 respirator if they have to go outside*".

The West Coast has the worst air quality on Earth right now, as nearly 100 active wildfires spew smoke, as of Sept. 12, 2020. Particulate matter from the smoke has made the air unhealthy to breathe all along the coast. A higher AQI indicates more pollutants in the air and a greater health hazard. "*The EPA considers any AQI above 150 to be unhealthy for all people*". Anything above 300 is considered a "**health warning of emergency conditions.**" The EPA does not make recommendations for AQI levels above **500**, since they're "*beyond index.*" But PurpleAir's monitors around Salem, Oregon, reported AQIs as high as **758** on September 11, 2020. Those levels are **NOT** comparable to some of the worst days for air quality in New Delhi (425 ppm), India, the world's most polluted city, according to the nonprofit Berkeley Earth.

"*None of PurpleAir's monitors at other locations across the globe was reporting AQIs anywhere near 400 on Friday morning, Sept. 11, 2020*". Some blazes are emitting so much smoke that they create their weather systems, and the haze has tinted the skies orange and red over San Francisco and other parts of the coast. The smoke was traveling so far, it reached the East Coast. When humans inhale these particles, they can penetrate deep into the lungs and even the bloodstream. Research has connected PM2.5 pollution to an increased risk of heart attack, stroke, and premature death. In healthy people, it can irritate the eyes and lungs and cause wheezing, coughs, or difficulty breathing (114a).

Plastic Pollution in the United States. More than a million tons a year of America's plastic trash isn't ending up where it should. "*The equivalent of as many as 1,300 plastic grocery bags per person (3.5 bags/day and person) is landing in places such as oceans and roadways*", according to a new study of U.S. plastic trash. In 2016, "*before several countries cracked down on imports of American waste*" the United States generated 46.3

million tons (42 million metric tons) of plastic waste, by far **the most in the world.** *If you took nearly 2.5 million tons of mismanaged plastic waste and dumped it on the White House lawn,"* it would pile as high as the Empire State Building," said co-author Jenna Jambeck, an environmental engineering professor at the University of Georgia. So some researchers from previous studies found that **America ranks as high as the third-worst ocean plastic polluter.** The study estimated that 560,000 to 1.6 million tons of U.S. plastic waste likely went into oceans *"We are facing a global crisis of far too much plastic waste."* The situation has been changing. China and other countries have become more restrictive about taking U.S. trash imports, and more plastic is ending up in landfills here. *"The best thing you can do environmentally is to produce no waste at all"*

Plastic Pellets are the most lasting pollution.

Some 30 miles north of Pittsburgh, Pennsylvania, in a township adjacent to a state forest, oil and gas giant **Royal Dutch Shell** is building a sprawling new plant to support what it sees as the future of its business: *"making millions of tons of new, virgin plastic."*

"The plant is just one of more than 300 new plastic facilities proposed or permitted for the US soon." Shell, along with Exxon, and Chevron, *"sees plastic as one avenue for growth"* as natural gas prices plummet. *"Industry aims to increase plastic feedstock production by at least 33% by 2025."* The Shell plant will rely on a process known as **"ethane cracking,"** where ethane gas, once seen as an unusable byproduct of gas extraction, can be molecularly **"cracked"**, its carbon and hydrogen atoms rearranged, *"to form ethylene, the main building block of plastic".*

The new facility will pump out 1.8 million tons of plastic each year. The vast majority of that plastic will likely not be recycled, it will exist virtually forever. The tiny plastic pellets, sometimes known as "nurdles," are a massive *source of plastic high pollutants. "With roughly **22,000 nurdles per pound** of plastic, the Shell plant produces an equivalent of 80 trillion nurdles per year".* In 2017, two shipping vessels collided, spilling 49 metric tons of pellets into the sea and coating 2,000 kilometers (1,243 miles) of South Africa's coastline with nurdles. *Researchers do know that pellets account for a whole lot of the world's total plastic pollution.* (115).

Oil Site in Texas Leaked Gases Uncontrollably for Months.

An oil well site in the Permian Basin owned by a bankrupt shale producer has spewed polluting gases into the atmosphere for **10 months,** despite being investigated by Texas regulators. Infrared video footage collected during multiple visits from November through September 2019 shows *"continuous intense and significant"* emissions from faulty valves and tank hatches at MDC Energy LLC PickPocket location in West Texas. They include *these intense emissions but are not limited to volatile organic compounds, methane, and hydrogen sulfide,"* Sharon Wilson, Earthworks' thermographer wrote. Those concerns are being compounded by a collapse in crude prices that's forced many producers into bankruptcy, and stop any activities.

In Texas, a major plastic-producing state, petrochemical company Formosa Plastics released millions of pellets into Lavaca Bay, a cove that washes into the Gulf of Mexico. In its wastewater permits, the Texas Department of Environmental Quality prohibits leaks of pellets, *"in other than trace amounts,"* which advocates say amounts to a pollution loophole. Scientists are worried about the lack of regulation to specifically address plastic pellet pollution. California is the only state with regulations to specifically control plastic pellet pollution. In other words, nothing has been done to regulate plastic pollution in the United States of America (116).

Pollution and Death.

Air pollution killed 476,000 newborns in 2019, with the biggest hotspots in *"India and Sub-Saharan Africa,"* nearly two-thirds of the deaths came from noxious fumes from cooking fuels. More than *"116,000 Indian infants died from air pollution in the first month of life"*, and the *"corresponding figure was 236,000 in Sub-Saharan Africa"*, according to the State of Global Air 2020. A growing body of evidence is linked to: *"mothers' exposures during pregnancy"* to air pollution with the increased risk of their infants being born too small (low birth weight) or too early (preterm birth). The new analysis estimated what percentage of those deaths came from ambient and *"household air pollution"*. Air pollution led to *"6.7 million deaths worldwide in 2019."*

Air pollution cuts life expectancy for every man, woman, and child on Earth by nearly two years. Poor air quality is *"the greatest risk to human health"*. In countries such as **India and Bangladesh,** air pollution was so severe that it now cuts average lifespans in some areas by nearly **a decade.** Some *"89% of the region's **650 million people** live in areas where air pollution"* exceeds the World Health Organization's recommended guidelines (117).

Half of US Deaths related to air pollution are from out-of-state emissions.

Not surprisingly, air pollution, like climate change, doesn't respect state borders. Over half of all air-quality-related early deaths in the U.S. are from pollution emissions from outside of the state. *"Pollution is even less local than we thought,"* study co-author Steven Barrett of the Massachusetts Institute of Technology told Reuters. Each year in the U.S., outdoor air pollution shortens the lives of about 100,000 people by one to two decades, MIT said. *"We used state-of-the-art modeling to estimate the number of air pollution-related deaths from each state that have been caused in every other state over the past 14 years,"* Barrett wrote in the Conversation. In 2005, for example, deaths caused by sulfur dioxide emitted by power plant smokestacks occurred in another state in more than 75% of cases. New York state was found to have the highest number of premature deaths, the study found (118).

Air Pollution Cost.

Air pollution costs from burning fossil fuels are generating economic losses of **$8 billion a day**, (compare with $5bn arm race) that's about 3.3% of global gross domestic product (GDP), or **$2.9 trillion per year**, according to a report from Greenpeace Southeast Asia and Center for Research on Energy and Clean Air. China, the U.S., and India bear the highest economic cost of soaring pollution, at an estimated $900 billion, $600 billion, and $150 billion a year, respectively. The earth could warm by 2C by 2050, cutting global GDP by 2.5% to 7.5%, Oxford Economics said in November 2019. PM 2.5 leads to the greatest health impact and cost due to increased work absences. *"In 2019, about 91% of the global population lived in places where levels of air pollution exceeded guidelines set by the World Health Organization" (119).*

Pollution & ESG

"A new kind of bond issued by companies that pollute has caught the attention of investors eager to fill their climate-friendly portfolios," according to the head of green bonds at JPMorgan Chase & Co., Paul O'Connor. Transition bonds are used by issuers to raise funding. These bonds are *set to play a key role in the assets that eventually live up to environmental, social and governance goals"* (ESG), said O'Connor. *"If we don't engage these sectors, not much is going to change."* With the demand for climate-friendly bonds, banks are advising borrowers to include sustainability in their funding. *"So-called brown companies (fossil fuel sectors) are also keen to embrace ESG. As a result,* the debt market started with *green bonds and more recently sustainability-linked securities."* Climate Bond Initiative last Oct. 2020, introduced a set of transition bonds. For instance, Etihad Airways sold $600 million in Islamic transition financing, with conditions of the **deal linked to the airline's emissions**. Also, the oil major was planning a *"very large transition bond sale"*. *"There's no getting around it"* (120).

Pollution Monitoring

Mitigating the effects of climate change and pollution is a global problem. Most communities don't have tools that can monitor emissions and pollutants they need for mitigation.

Aclima, founded by Davida Herzl, is looking to solve that problem.

"The company has all the technology to make all measurements for GHG and pollutants. The co. set all data together and bring that into a back end," Herzl said. Aclima is working in partnerships with Google, which captures quality data alongside geographic information. *"The interconnected crises of climate change, public health, and environmental justice urgently require lasting solutions."* Measurement will play a key role in shaping solutions and tracking progress. With this coalition of investors, we're expanding

our capacity to support new and existing customers and partners taking bold climate action.

Governments and industries will need Aclima's data to reduce emissions. An investor in Aclima is Microsoft, which has backed investments from the Microsoft Climate Innovation Fund (122).

Pollution, Cynicism & Oil Industry.

The Environmental Protection Agency (EPA) said on March 24. 2020, that it's reviewing a "*request from the oil and gas industry **to ease** enforcement of hazardous air and water pollution*" and other regulatory issues during the coronavirus pandemic. This proposal is against public health and the environment. "*The oil and gas trade group is complaining of staff shortage issues during the outbreak, saying worker shortages could make compliance with a range of regulations difficult.* EPA spokesman Corry Schiermeyer said the agency was evaluating the request. EPA policy explicitly prohibits environmental and public health laws. The fossil fuels companies' request is "*alarming*" and "*wildly overbroad.*" The oil and gas industry already is a major contributor to hazardous pollutants through air and water emissions.

Trump's EPA rewrote the rules, in Oct. 2020, and President Trump signed an executive order to abolish all EPA regulations. Death on climate change actions. The rule is one of nearly 100 environmental rollbacks Trump has signed, "*to loosen regulations on everything from air and water quality to wildlife* (121).

4

ARE FOSSIL FUELS
IN THE FALL?

Before looking in detail for the direction where fossil fuel might go, let see an analysis of the direction past the pandemic crisis. The coronavirus pandemic wiped out millions of barrels in fuel demand. This demand is back to the pre-pandemic crisis, as of May 2021. *Chinese refiners have been importing oil like there is no tomorrow"*. The economy is recovering, they will need more oil, this includes Europe. The big oil company BP Plc. says the *"era of oil-demand growth Is over"* That will prove (mid. 2021) otherwise. High heads of big fossil fuels say that oil consumption will see decades of growth. They have described it as the only commodity that can satisfy the demands of an increasing global population and expanding middle class.

Transport is a huge market for crude oil and is not going away. As of March 15, 2021, the oil price reached the previous pandemic price ($64/barrel). Oil businesses return as seen in the above graph, we need to look at "the business-as-usual". The industry's primary demand sources will continue to be there even in 2050 (123).

The Big Fossil Fuels industry, as shown in the above chapter, had and has invested billions and billions in existing and new fuel sources. They will not easily give up on their investments, because some governments require them.

4.1. BANKS, INVESTMENTS

4.1.1. E.S.G. ENVIRONMENTAL SOCIAL AND GOVERNANCE.

Before we present the financial aspects of Global Warming, let's briefly see how much it will cost to try to slow down Climate Change. In 2019, Time reported that U.N. climate scientists estimated that global warming could be attenuated with $300 billion. (Note: These scientists in particular were looking at tackling the agricultural aspect of climate change, chiefly desertification). The big picture of climate change is much pricier: The International Renewable Energy Agency says *"$750 billion a year is needed in renewables over a decade." Total of $7.5 trillion/decade.* Then you've got the cost of *"carbon capturing and storage ($2.5 trillion), $ 2.7 trillion for biofuels per year,"* and the list goes on. In total, we are looking at *"at least $50 trillion"* to attenuate climate change. It's a scary number, huffs. As shown in the conclusion (chapter 7), **more than $5 trillion a year spending** of the entire humanity will be the minimum spending to only slow down global warming. That will be the real cost of luxury to consume fossil fuel in excess. There's a *"green investing trend"* on Wall Street. *ESG* investing is a more than $30 trillion market. *"Environmental"* covers corporate initiatives on issues like *"climate change"*, carbon emissions, and pollution. *"Social"* covers the company's treatment of key stakeholders, ranging from employees to consumers. *"Governance"* covers managerial aspects like board diversity, executive pay, and accounting. Investors are looking at companies that prioritize ESG. Mutual funds and *"ETFs (exchange-traded funds)"* offer popular ways of buying considerable ESG-focused companies. (EFT stands for Electronic Transfer Funds too). ESG investing at large has exploded in popularity, and regulators are ramping up scrutiny. The United Nations has already created a *"Principles for Responsible Investment (PRI)"* establishing an international standard for defining ESG investments. In December of the last year 2019, the *"Securities and Exchange Commission (SEC)"* said it would be investigating criteria by which *"investment funds truly abide by the UN PRI"* (124).

4.1.2. BANKS

Banks are far more exposed to climate change than they're disclosing. The financial sector may face losses from inter-bank lending and *"fire sales" of "distressed high-carbon assets"*. JPMorgan Chase agrees with Paris climate pledges. The US bank's exposure is so significant that it could trigger a financial crisis, and significant climate risk, (more than 100 billion dollars in losses).

Banks and investors are more demanding transition and green bonds and loans to the fossil-fuel sector. *"While there is growing interest in acting on climate risks from U.S. financial regulators, banks have an opportunity to act on these risks now, without waiting for the policy or regulatory change. They can help prevent the next financial crisis.* Banking shares, boosted by earlier COVID-19 stimulus and strong capital reserves, have seen a rebound (125).

4.2. IS THIS THE BEGINNING OF THE END OF FOSSIL FUELS?

The eventual death of oil and thermal coal won't come from environmentalists, or even directly from renewable energy, *"it will come when big banks decide to stop financing it, rendering it 'unbankable and unsustainable"*. That's exactly what Goldman Sachs has just done. *"Goldman Sachs"* is the first big U.S. bank *"to rule out financing new oil exploration or drilling in the Arctic, as well as new thermal coal mines anywhere in the world"*. The bank's latest environmental policy declares climate change as one of the *"most significant environmental challenges of the 21ˢᵗ century"*, and has pledged to help its clients manage climate impacts, including through the sale of weather-related catastrophe bonds. The bank will invest $750 billion over the next decade in climate transition. *"The so-called Big Six banks"* tend to move in lockstep, and the other five will feel high pressure to forgoing financing fossil fuel projects. Investing in the *"Arctic is most climate risk-able, with the region warming at nearly twice the global average rate"*. More than a dozen of the world's largest banks including UniCredit, Royal Bank of Scotland, and Barclays have pledged to stop drilling in the Arctic. For instance, several green groups from the U.S., the UK, and the Netherlands asked the world's largest banks not to underwrite Saudi Aramco's IPO (37).

It is for the first time a *large U.S. bank has cut off financing for the entire fossil fuel sector."* Goldman Sachs's updated policy shows that *"US banks can draw red lines on oil and gas." "The coal industry is dying,"* a slow and painful death. The same is true for Arctic oil (126).

The World's Bank Plans to Tackle Climate Change. *"JP Morgan Chase"* Here I present a model of **ACTION** for the big banks in tackling the *"Climate Crisis"* as of Oct. 16th, 2020 (126). Other banks, investors, and stock manager companies have to follow in. *"Climate change is among the most urgent problems facing humanity".* Fossil fuel resources which *"supply 80% of the world's energy",* will continue to exist and do good business. JPMorgan Chase will focus on three sectors in particular:

1. oil and gas, 2. electric power, and 3. automotive manufacturing;" because they collectively generate a significant share of global emissions. The bank shall establish sector-by-sector intermediate emission targets for 2030. The intermediate policy is one of the best to reduce the emission of GHG and tackle the climate crisis. There will be skeptics, to reach carbon neutral in 2021, we know, and those who choose to delay. *"But now is the time to act, to forge an economy that has sustainability and efficiency at its core and reduces pressure on the planet's resources." (Daniel Pinto is co-president of JPMorgan Chase and CEO of its Corporate & Investment Bank. Ashley Bacon is the chief risk officer of JPMorgan Chase, the authors of this article)"* (126).

4.3. INVESTMENTS AND FINANCING

Blockchain empowers people to buy from companies that share their values, giving consumers the info needed "to contribute to an inclusive green economy."

ESG"- *the new shorthand for environmentally-minded corporate do-gooding",* a new definition that one presented in paragraph 4.2. *Companies must reduce their emissions, using the blockchain to identify who is emitting what, and getting credit for cutting your emissions."* For instance, EV tailpipe emission is zero (if the electricity used to charge the battery is green) but the battery and body of the car's carbon footprint are covered after driving the car 60,000 miles. Only the electric battery involves a huge carbon footprint, half of the rest of the car's body. An expensive battery from a manufacturer who used green energy to produce it may save 5 years of the carbon neutrality footprint.

Accountability. The carbon accounting issue was presented by *"Hyperledger", a climate action group. The most climate actions nowadays are companies first fossil fuels companies and then all production companies using gray electricity.* The target of the project is to *"create a global climate accounting mechanism that can be compatible with the Paris Agreement."*

Digital transformation is used for a good climate accounting system by blockchain. *"You have to think about a global climate internet. Hyperledger has become a place to apply all these issues (127).* Investors, all over the world, have the tools required to access any company in any field to make sustainable investments, which includes climate change via climate planetary internet. Following I present the most significant examples of sustainable investment, out of hundreds of cases.

Eyes in The Sky

How we put numbers on what climate change means for countries, companies over the next 30 years. This involves special finance via satellites. So, analysts acquire satellite imagery, filter them using algorithms, and then use the results to control how climate change could affect anything from a single factory to an entire economy. Asset managers tracing market debt are the first to explore the possibilities. A country with intense hurricanes or heat waves can raise the finances of countries that depend on agricultural exports.

"A big question is, what happens when particular companies, assets, and entire countries are identified as being at risk? Are those assets sold to markets and buyers who have the same visibility of that risk?" In Britain, new spatial finance start-ups include Oxford Earth Observation and Sust Global. Astraea Inc. in Virginia, USA, extracts data from 1,500 earth-observing satellites. The company is partnered with Caldecott's spatial finance initiative to create a database of all the world's cement factories, giving investors possibilities to push the *"most polluting operators"* to clean up their act (128).

Finance Industry and Sustainability Rules

"The accelerating climate crisis is increasing pressure on regulators and corporate executives to set universal standards for the burgeoning sustainable finance industry".

Sustainable finance has already expanded to a large area of the economy. **But there's a problem**: *the absence of consistent guidelines and definitions with which players can examine markets," including carbon credits, ESG funds, and green bonds."*

The regulators are working together to address these challenges. *"The rules to be standardized in one single measurement, regulation, which would make it easy to use."*

"Former Bank of England Governor Mark Carney said one of his main focuses is achieving clarity in the market for carbon offsets, which can be built into something much larger to help achieve net-zero global emission targets". More about carbon offset in Chapter 6 (129).

The $30 Trillion Economic Trend.

Sustainable ESG investing, already a *"$30-trillion megatrend"*, is calling all the shots in a world where big money has realized one critical truth: *"This is a game of survival, and the key is mitigating risk" (making money).* Big money goes where they can grow. From here they go to profitable sustainability. The $30 trillion marks was reached in the first half of 2019, according to the Global Sustainable Investment Alliance (GSIA).

Any company will love to catch on to the $30 trillion ESG trend, including **Big Oil and Gas** supermajors. French oil company **Total** has contributed to each of the United Nations' Sustainable Development Goals. This same trend is while renewable energy giants have done so well in recent years like **NextEra,** a dedicated wind & solar energy. It's no surprise that it has received some love from the 'millennial dollar. **Alphabet and Facebook** are doing their part in Big Tech's renewable push. **(FB)** made progress *"running on 100% renewable energy by the end of 2020,"* and working to build more weather-efficient data centers (130).

Why BlackRock Is Going Green. BlackRock *has* announced that it *"would exit investments* with *high sustainability risks".* The BlackRock CEO's letter to its customers shows that investment policies do change, and that is not good news for the fossil fuel industry. Big hedge funds like BlackRock and Goldman Sachs have stashed billions aside to invest in ESG, but they can't find enough sustainable stocks to invest their billions. Wall Street is benefiting from these ESG portfolio hedge funds. (128).

Climate Action 100+ aims to cut greenhouse gas emissions for investors.

"Saudi Aramco, the world's largest oil producer, is on investment group list of target companies." Climate Action 100+ (CA100+), the *"450 members, and manage more than*

$47 trillion in assets," try to engage with big emitters of GHG. The group achieved success with Royal Dutch Shell and BP, which in 2020 announced a long-term approach in response to climate change. *CA100+ said it had added big companies around the planet to achieve ESG criteria* (131).

Goldman and Renewables vs Oil and Gas

"Spending on "renewable power is set to overtake oil and gas drilling for the first time next year 2021, as green energy affords a $16 trillion (maybe more)investment through 2030." Goldman Sachs Group Inc. *"Clean energy may attract $1-$2 trillion a year"* in infrastructure investment and *"create 15-20 million jobs globally.".* Renewable power will become the largest area of spending in the energy industry in 2021 (130).

Microsoft Makes First Climate Fund Investment.

Microsoft Corp. allocated $1 billion climate fund investment, in venture capital firm Energy Impact Partners. *"The software also joined with Nike Inc., Starbucks Corp., Unilever NV, and Danone SA"* for *"slashing carbon emissions".* Microsoft announced in January 2020 that it intends to be carbon negative, by 2030. Microsoft will partner with Sol Systems, a renewable energy developer to invest in 500 Megawatts of solar energy. This solar plant can power about 95,000 homes. The partner group included many big international corporations (132).

Bottom Line For Financial Regulation.

President Joe Biden plans to use every tool at his disposal in the fight against climate change, including financial regulation, much needed in the USA, as well as internationally. Mandating that public companies and investment firms quantify and disclose climate risks, is a bold step that could make ESG data as commonplace as sales and profit. *"The new administration in Washington has given a renewed sense of urgency para climate crisis."* The Treasury Department is adding a *"climate czar."* In other words, *"banks and insurers could be subject to climate stress tests, and certainly require a policy."* Climate change is a crisis of such monumental significance that using financial regulations as a lever to advance environmental policy is less extreme than it sounds (USA SEC).

4.4. FINANCING CLIMATE CRISIS

Amazon CEO Jeff Bezos Pledges $10 Billion to Fight Climate Change

Jeff Bezos is committing $10 billion of his own money to fight climate change, creating the *"Bezos Earth Fund."* The Amazon CEO said: humans can save the Earth by *"using an inclusive approach that combines the efforts and resources of all stakeholders. We can save Earth. I'm committing $10 billion to start and will begin issuing grants. Earth is the one thing we all have in common, let's protect it, together."* (133).

Climate Change Is a Financial Crisis, Too.

Inaction by governments is leaving the country exposed to a devastating crisis.

"Financial institutions, from banks to insurers, asset managers, face climate-related risks. that goes far beyond the issue of social responsibility.

Natural disasters and forced migration, or decisive moves to transform energy use, could end up facing trillions of dollars in cumulative losses. These will be in the form of defaulted mortgages in flooded areas, investments in regions that become uninhabitable, or nonperforming loans to shuttered coal-fired power plants. In the last three years, the six largest U.S. banks provided more than $700 billion to the fossil-fuel industry. As of 2016, large insurers held $528 billion in fossil fuel investments. By *one estimate, some $200 billion in capital must be reallocated* in each of the next 40 years. This means by 2050 a total of $6 trillion to limit global warming. Can we save it? The Financial Stability Oversight Council, charged with monitoring and addressing systemic risks, must mitigate climate-related risks outside the banking system. *"Science is not a counterparty that can be negotiated with, the planet is not a contract that can be restructured, and there is no bailout for a climate catastrophe"* (134).

£30bn Pension Fund, threatens to sack money managers who ignore the climate crisis

A £30bn ($39.3bn) pension fund has said it will vote down boards and sack asset managers whose investments fail to meet climate targets. The Brunel Pension Partnership said in January 2020, *"its money managers must move towards carbon-neutral investments by 2022 or risk being axed"*. Investment managers were too short-term focused and not disclosing risk by climate change. *"We need to systematically change the investment industry,"* The Brunel Pension Partnership said (135).

How California's Wildfires Could Spark a Financial Crisis.

Wildfires are among the sparks from climate change that could ignite a US financial crisis by damaging home values, state tourism, and local government budgets, and more. Those effects could set off financial defaults and market disruptions, undermining the US economy and sparking a crisis. Here's how: According to the report, produced by a 35-member panel for the Commodities Futures Trading Commission, devastating fire may change the entire financial market.

-CalFire, California's fire-fighting agency, says about *"3 million of the state's 12 million homes are at high risk from wildfires."*

-After 2018, California's worst fire season in terms of loss of life and property, some insurers balked at renewing homeowner policies. Owners may need to find new pricey issuers.

"You can tell the same story in terms of sea-level rise and flooding and more intense storms and their impact on residential real estate value."

-Lower home values reduce cities' real estate tax revenue and impair their ability to repay city employees.

-Climate catastrophes can make investors aware of risks not priced into markets.

The same financial risk assumption can be made referring to extreme floods, droughts, and hurricanes, like weather phenomena occurring more frequently, and intensively (136).

Wealthy Countries Financing Developing Countries.

Wealthy countries have increased financing to help developing countries. But this goal has been missed in 2020 to pay $100 billion. Developed countries agreed at the United Nations in 2009 to contribute $100 billion each year by 2020. The max amount paid was in 2018, $14.6 billion. European countries increased their contribution in 2019 by $21.9 billion (138).

UBS Chairman: Climate Change Occurring at 'Astonishing' Speed

The weather distracting events from the last decade, as well as scientific data showing that it is time for banks and investors to think twice about the climate crisis.

UBS bank has been exposed to the data on climate change for 30 years.

"If you look at the current research you know this: climate change is for real. The speed at which the environment is changing is astonishing. The UBS chairman's comments came at UBS's first-ever ESG and Sustainability Symposium in London in September 2019. Weber called on investors and finance professionals to make tackling climate change a key priority. *"Financial markets are enormously effective in moving capital where it is needed most,"* he said. *Financial markets have to mobilize capital and align capital allocations more strongly with sustainability?* (137).

ENERGY

5.1. PRIMARY ENERGIES

There is a distinction between energy and electricity. Energy includes all forms of combustion material for heating and produces electricity. The heating process includes houses, industry (steel, aluminum, cement, etc.) transportation, agriculture, and other sectors. The use of all these forms of energy emits greenhouse gases (GHG). For the year 2019, for which BP (British Petroleum) provided a summarized data on global energy. The primary energy to produce electricity from fossil fuels is presented in the following chart. (139).

Global fossil fuel share of primary energy:

Fuel Consumption (ExaJoules EJ) Share of primary energy.

Oil 193.0 31.1 %

Natural Gas 141.5 24.2 %

Coal 167.9 27.0 %

Renewable 29.0 5.0 %

Hydro 37.6 6.4 %

Nuclear 24.9 4.3 %

Others (like biofuel) 11.6 2.0 %

Total 605.5 100.0 %

Note: Exa = 10 at 18 power.

Fossil fuel energy represents 82.3 %, versus renewable including nuclear and hydro 15.7 %, which is 5 times less than fossil fuel share. To reach net-zero emission by 2050, or earlier, is a very long, hard, and costly, achievement. Compared with the previous year 2018, electricity generation grew by 1.3%, around half of 10 years on average. China produced more than 90% of the net global growth, with an annual rate of 8.5%. Renewable energy has the largest increments, followed by natural gas, hydro and nuclear energy.

-Coal share of the generation fell 1.5 % to 36.4 % (including coal used for heating). Coal consumption decreased by 0.6 %, and its share in primary energy fell to 27%. Coal increased consumption was driven by China and Indonesia. OECD consumption fell to its lowest level. Coal global production rose by 1.5%, with China and Indonesia providing the only significant increase. The largest decline comes from the USA and Europe.

-Oil consumption grew by 0.9 million barrels per day (mb/d), or 0.9%. Demand for all fuel rose by 1.1million b/d and topped 100. million b/d for the first time. Oil consumption growth was led by China 680,000. b/d, and other emerging economies, while demand for OECD fell by 290,000. b/d. Global oil production fell by 60,000. b/d, as strong growth in the US output 1.7 million b/d was more than offset by a decline in OPEC production of -2.0 million b/d., Iran 1.3 million b/d), Venezuela 560,000. b/d, and Saudia Arabia 430,000. b/d.

-Natural gas consumption increased by 2%, 78 billion cubic meters, - bmc, below the growth seen in 2018 of 5.3%. That led to a growth in primary energy to a record high of 24.2%. The increase in gas demand was driven by the US 27. bmc, and China 24. bmc. The gas product grew by 132 bmc, with the US accounting for almost ⅔ of the increase 85 bmc. Australia and China were also key contributors to growth 23 bmc and 16 bmc respectively.

Liquid Natural Gas (LNG) supply growth was led by the US 19bmc, and Russia 14 bmc, with the most supply heading to Europe 49 bmc.

-Renewable energy including biofuel, had a record increase of 3.2 EJ. Wind provided 1.4 EJ, followed by solar with 1.2 EJ. By countries, China had an increase of 0.8 EJ, followed by the US 0.3 EJ and Japan 0.2 EJ.

-Hydro energy consumption growth below average 0.8 %, with growth led by China 0.6 EJ, Turkey 0.3 EJ, and India0.2 EJ.

   -Nuclear consumption rose by 3.2 % (0.8 EJ). China 0.5 EJ and Japan 0.1EJ, produced the largest increments. The increase in gas demand was driven by the USA and China, while Russia and Japan saw the largest decrease. These are a few numbers for the 2019 energy development. The energy data for 2020, which will be released in mid-2021, will strongly be different from that of 2019 (139). There is so much data about ENERGY, if we look at this issue through all aspects, and try to select the most important of them: economically, reducing emission, the transition to renewable energies, storage of energy, social, political, technological aspects, *"then all looks like navigating through the Bermuda Triangle."* To include in this book all aspects shall take hundreds of pages, so I will choose more important aspects for slowing down global warming. In the next part of the chapter, several times will appear the notions of watts, kilowatts, megawatts, and gigawatts. (1kW =1000W; 1MW =1000KW = 10e6W; 1GW 1000MW =10e9W. e6 is 10 at 6 power).

Energy Efficiency

   One of the best and most equitable ways to address climate change is to push "energy efficiency" **into every corner of our economy** (industry, housing, transportation, agriculture, and so on), to reduce consumption. According to federal research, households earning less than twice the federal poverty level, roughly less than $50,000 for a family of four, spend an average of 16% of their income on energy costs. Households earning above $100,000 spend just 3.5%. Energy efficiency is an undervalued solution. Discussion about environmental justice, and climate policy, should start with energy efficiency. Consider that one efficiency program alone, the Department of Energy's minimum efficiency standards for common appliances, saves the average household more than $500 per year. *"The International Energy Agency projects using existing technology can account for nearly half of the emissions reductions needed to meet the goals of the Paris climate accord. Energy efficiency also offers an enormous opportunity for job creation"*. Already, the energy efficiency sector is one of the largest energy workforces in America with more than 2 million employees, 12 times more than the coal industry, and nearly seven times that of wind and solar combined. Most of those jobs are in construction and manufacturing, and 80% of efficient companies are small businesses with fewer than 20 employees. It's time for Washington to adopt more urgency by pulling policy levers at its disposal and making strategic investments. For example, a program to help some 35,000 low-income households per year upgrade their homes to be more energy-efficient, is underfunded. Similarly, we should help our small businesses improve efficiency with federal grants matching existing utility incentives. Energy bills, after all, are often the second-highest cost of owning a home behind the mortgage. Energy efficiency is also good business, it reduces strain on the grid and strengthens economic productivity and competitiveness. (Used watts for illuminating an average-size home: 1Megawatt = 250 homes: 1Gigawatt = 250,000 homes).

5.2. RENEWABLE ENERGY

5.2.1. SOLAR AND WIND ENERGY

Corporate America, Renewable Energy. At the current moment, what is blocking renewable transitions in the U.S, *"is politics." (*Note: This was under the Trump administration). When the new President of the United States of America takes office, all renewable energy issues will go in the right direction stopping the toxic lobbying in Congress. And in this field, there are a lot of regulations to be fixed. I have to skip this issue and keep with the subject of the chapter. To change the course in renewable energy, environmental groups, and influential companies, to push state legislatures to adopt new policies. Examples of such companies are *tech companies like Google, Apple, Salesforce, Facebook, and some large retailers, Walmart, IKEA, and they have a lot of power over legislators"* (140).

Today The Price Of Clean Power For America.

The estimated cost of reducing carbon emissions is falling rapidly. The transition to cleaner energy is much less costly today than it used to be. *"Three forces are changing the math."*

First, renewable power costs are dropping so fast. Technology, innovation, and lower capital costs drive the low cost of renewable energy.

Second, the cost of storing renewable energy is also falling. Wind and solar are intermittent, so they require storage or support from other sources like gas power. Storage technologies are evolving rapidly and costs are plummeting.

Third, and crucially, the old power plants need to be removed and replaced. So now is an opportune moment to jump to new green energy plants. *Pricing carbon"* is detrimental for new sources of energy, these advantages and others like subsidies will provide in choosing green energy. *"For $20 a person per year, it's possible to eliminate net carbon emissions from our power grid within three decades".* So, the United States shall reach net zero by 2050. (141).

Amazon Becomes The Largest Corporate Buyer Of Wind And Solar

Amazon has announced *"26 new utility-scale wind and solar energy projects"*, making it the largest-ever corporate purchaser of renewable energy. The company has now invested in 6.5 Gigawatts of wind and solar projects in total. They will supply their operations with more than 18 million megawatt-hours of renewable energy annually, enough to power 1.7 million U.S. homes for one year. *"This is just one of the many steps we're taking that will help us meet our Climate Pledge. Amazon aims to have net-zero carbon emissions by 2040".* This new Pledge *has 31 signatories, including Unilever, Verizon, Siemens, and Microsoft"* (142).

Renewable Power In 2020

Renewable energy grew in 2020 and will represent about 90% of the total power capacity added for the year," according to the International Energy Agency (IEA).

A surge in new projects from China and the U.S., which will account for almost *"200 Gigawatts around the world",* according to the IEA. European Union will also jump in and add 10% of renewable capacity. Governments provide continuation of subsidy programs, solar and wind additions could jump by another 25% by 2022. This way solar energy *"could reach a record 150 gigawatts by 2022, a 40% increase in just three years"* (143).

U.S. Solar in 2020

Solar installations are expected to soar 43% in 2020.

The improved outlook reflects robust demand from utilities seeking to meet carbon-reduction goals and a rebound in demand for home solar systems, thanks in part to declining costs for the technology. The sector shall install more than 19 Gigawatts of solar, enough to power more than 4.6 million homes. Last year (2019) it installed 13.3 GW of capacity. Even so, *just 3% of U.S. electricity is now generated by the sun, in a decade should jump to more than 20%"* (140).

Scottish Water Start Installation of New Low-Carbon Project.

Scottish publicly owned Water started to install a 2 million pounds solar power and battery storage facilities near Perth's Wastewater Treatment Works at Sleepless Inch on the River Tay, Scotland. Renewable energy is expected *"to provide 25% of the electricity needed to treat water"*. It will be one of the *"first Scottish Water solar projects to include battery storage*. Will be complete in 2021. By using this solar power on-site (no grid required) *carbon footprint will be cut by around 160 tons per year (tpy) of CO2.* In addition, Scottish Water Horizons' project will also include the installation of electric vehicle charging facilities of the 1600-vehicle fleet. The company made the acquisition of needed equipment. (145).

"If this solar power energy can be achieved in Scotland, not in the Sahara Desert, the rest of the world needs to look at this developed project." These examples of projects show that there is the technology and means of achieving carbon neutrality by corporations, earlier than 2050.

BP Joins Up With Equinor For $1.1 Billion Offshore Wind Partnership.

BP PLC and Equinor on Sept. 10, 2020, will develop offshore wind projects in the US. BP will buy a 50% interest in Equinor's US wind assets, for $1.1 billion.

The partnership comes just a month after *"BP announced to increase annual low carbon investment 10-fold to around $5 billion a year.* Also, BP will increase green energy power from 2.5 GW in 2019, to 50GW by 2030.

TOTAL Partners Ignis For Renewable Operations In Spain.

TOTAL SE announced that with Spain-based developer Ignis to develop 3.3 GW of solar projects. The project will start in 2022 and all the projects will start production from 2025.

"Spain's current ambition is to generate 70% of electricity from renewables by 2030 and 100% by 2050." (146).

Australia's Ambitious $16 Billion Solar Project, World's Biggest.

One of the most ambitious renewable energy projects in Australia– ASEAN Power Link. There will be: *"the world's largest solar farm, the largest battery, and the longest undersea electricity cable."* The 10GW solar farm would cover 30,000 *acres (12,500. hectares) in Australia's Northern Territory."* The farm is associated with a *"30GWh battery storage facility."* The facility will transfer to a *"3,700 km 2.2 GW undersea power line"* to Singapore. Sun Cable, a Singapore-based company behind the proposed $16 billion projects (147).

Green energy. Due to public and investor pressure, big oil and gas companies pledge net-zero by 2050. The only transition from fossil fuel to clean resources is a difficult issue (148).

The True Cost Of The Global Energy Transition.

"*$15 trillion: this is the amount of money to be invested in new power capacity globally over the next three decades"*. **Note**: Based on the last data on the climate crisis, this estimate is **under-evaluated**. Most of this, *"80% will be poured into renewables"*. Bloomberg NEF, *"also estimate that between 2020 and 2050, another $14 trillion will be indeed for the grid."* The materials for solar panels and wind plants are carbon expensive, and also battery storage has a high level of carbon footprint. Besides these, there are other inconveniences and burdens to be overcome. For instance, by 2050 up to 500GW in storage is necessary. Is a huge amount of rare metals, like lithium-ion for batteries. The good news is that in time, with innovation battery prices will go down. Rare metal might cost another trillion dollars for the next 3 decades. Green energy material can not be recycled 100%, which creates environmental issues, adding more cost for transition. During the transition period, society must suck billions of tons of GHG from the atmosphere to keep the balance between release and absorb greenhouse gases. Also, *"there appear to be other, half-hidden costs that are not just financial but also social and environmental."* (149).

Commercial Energy Storage In The US

The energy storage industry was *"exploding"* as grid-connected energy storage had increased significantly around the globe, with an annual growth rate of *"74% worldwide in the years 2013 to 2018,"* according to Wood Mackenzie's analysis in April 2020. **China** dominates the market, as its *"energy storage capacity is projected to increase from 489MW or 843 (MWh) in 2017 to 12.5 GW or 32.1GWh in 2024,"* an increase *of 25 times."*

The United States is also set to significantly increase its capacity in the coming years and with China *"set to acquire over 54% of the market by 2024"* Climate change creates *a new approach to energy and, very important, energy autonomy"*. One example is the Grid Problems created by the fire in California and Vortex in Texas, presented in Chapter 2. All these changes in energy power, and during the transition from fossil fuel to renewable arrived at *a "new energy order."*

as the World Economic Forum advocated (150).

Dutch Offshore Wind Farm Ready To Deliver.

Danish energy firm Orsted said in 2020, it had finished building *"the largest offshore wind farm of the Netherlands,"* 752 megawatts (MW) wind farm is the *"second-largest in the world. Can* power the equivalent of around *"a million Dutch households."* The Dutch government approved building wind farms of 2800 MW capacity in the North Sea. In the next 5 years, 6100 MW of new wind power will be built. The Dutch aim to get 40% of all their electricity from wind farms by 2030 (140). There are hundreds of projects on the field, and design tables for renewable energy, and storage facilities, which can and will concur with carbon neutrality. I stop here with the exemplification, hope the most significant aspects have been reached (152).

5.2.2. HYDROGEN AS ENERGY

General Notes: Hydrogen is the most potent source of green energy to replace oil and natural gas. There is no free hydrogen in nature, it has to be obtained from water, (green hydrogen), coal (gray hydrogen), and gas (blue hydrogen). Hydrogen is used as rocket fuel, and to produce ammonia for fertilizers. *"Blue hydrogen has lower carbon emissions and is cleaner than most fossil fuels".* Carbon emitted can be captured and sequestrated. *"The best is to extract hydrogen from water using electrolysis powered by green energy".* To be competitive *"green hydrogen",* has to reach a cost of $1.5 per kg.

At least *"$150 billion worth of green hydrogen projects announced globally in 2020.* 70GW of such projects are in development, which could require a $250 billion investment by 2040.

Industries that use coal, oil, and gas as raw materials, such as in steelmaking, cement, chemicals, and fertilizers, and shipping, and aviation. The key costs to meet the sub-$1.50 target are the cost of wind, solar power, and electrolyzers. The cost of wind and solar fell to 50%, and electrolyzers dropped to 75%. to meet that target. To ship liquid hydrogen long-distance it has to be chilled to minus 253C. It is much easier to *"ship hydrogen in the form of liquid ammonia, which needs to be chilled to only minus 33C."*

The European Union has plans to invest up to *"470 billion euros ($570 billion) of investment in green hydrogen by 2050".* U.S. President-elect Joe

Biden wants to fund research to make green hydrogen costs competitive on the market in 10 years (153).

Blue Ammonia

"Ammonia is" a chemical compound that contains *"three hydrogen molecules and one nitrogen molecule, and through burning doesn't release carbon dioxide.* The blue ammonia may be obtained from coal and gas, using capture carbon systems for sequestering the CO2 during production. For the first time *blue ammonia from Saudi Arabia was shipped to Japan"*. This shipment consisted of 40 tons of *"high-grade"* blue ammonia that will generate power in Japan, with zero emission. Big Oil's most profitable business is no longer oil, is ammonia (153a).

Science and Technology to Produce Competitive Hydrogen.

"The U.S. Department of Energy (DOE) is investing in research to obtain green hydrogen using nuclear power." This is a challenge for both nuclear energy and carbon-free hydrogen production. *Hydrogen obtained this way will be the energy of the future, with profitability, and grid reliability, zero-emission, and a large scale of usage. The hydrogen power plants produce intense energy used in metal, cement, petrochemical, and other heavy power consumptions.* (154).

For instance**,** NuScale's small nuclear module performed effective catalysts for hydrogen.

With the 25% increase in power output of NuScale Power Module, a 250 MWt NuScale module can produce 2,053 kg/hour of hydrogen, equivalent to 50 metric tons per day." This performance must be taken into consideration. Small nuclear energy is on the rise around the world, but NuScale has drawn the most attention. NuScale's module is *"heating water and then zapping the hydrogen out of the superheated steam."* This module *"has improved containment and safety outcomes, due to advanced design in these reactor units'.'*

The US DOE awarded grants for a project called H2-Orange, in collaboration with Clemson University. The project includes studies of hydrogen production, storage, and co-firing with natural gas, to obtain green hydrogen. Hydrogen can store large quantities of energy more efficiently than lithium-ion batteries. This is a collaboration between Siemens Energy, Clemson University, Duke Energy (155).

Hydrogen From Dirty Coal. A Japanese-Australian partner has begun producing hydrogen from brown coal. Liquefied hydrogen can be produced on a large scale and exported. The goal is to create an international supply chain for liquefied hydrogen. *"The project has the potential to produce and export hydrogen."* Australia has a big piece of the pie in the global liquefied natural gas (LNG) trade. The new facility is run by Japan Kawasaki Heavy Industries and is located in the state of Victoria Australia, which generates a *"quarter of the world's known brown coal"* reserves. Japan's target for 2050 of *"20 million tonnes"*, is equivalent to *"about 40% of its current power generation"*. The process of obtaining hydrogen also yields carbon dioxide and other gases. The carbon dioxide would be buried. H*ydrogen produced from coal, using CCS is half to one-third the cost of producing regular green hydrogen"*.

Another way to make hydrogen cheap. Scientists at the Pacific Northwest National Laboratory (PNNL) have found that molybdenum-phosphide (MoP) catalyst with wastewater in a small reactor called a microbial electrolysis cell (MEC), worked better than platinum. Platinum is used as catalysts, but very expensive rare minerals. The MoP catalyst excelled at working with seawater. The switch of catalyst on the process of electrolysis, and use of seawater can lower the hydrogen price to $2/kg., a very competitive fuel.

The green hydrogen hype. Fossil fuel or not, **big** companies are making a "trillion-dollar bet" on a hydrogen-fueled future. Around the globe, hydrogen could eliminate 25% of GHG emissions. Innovation and technology have opened the field to produce green hydrogen on a cost commercial scale. Wind and solar power compete economically with fossil fuels. The excess renewable energy can be used to produce hydrogen, which may be stored. Big industries such as petrochemicals, *"use hydrogen to make* *ammonia".* Any vehicle running on a gas-powered engine could eventually have a *"fuel-cell system powering it instead."*

Nuclear plant to produce commercial green hydrogen.

"It will not be many years" before many of these projects *"reach an industrial scale,"* and get industrial *"hydrogen production."* (156).

Storing hydrogen in caves. Salt mines can store large quantities of hydrogen for energy.

It is convenient, and it solves a huge structural problem containing huge amounts of hydrogen. Green fuels may be used at any time to produce electricity or as engine fuel.

Spain Invests $10.5 Billion in Green Hydrogen.

Spain to be on par with France and Germany in seeking a greener fuel for heavy industry. The government shall build 4GW of green hydrogen capacity by 2030. *"The project would cost 8.9 billion euros ($10.5 billion)."Spain, by taking advantage of the high potential of generating renewable power at very competitive prices, may use it for hydrogen production".*

Spain production would represent 10% of the EU's target, which is for 40GW by 2030. Spain has 61.2GW of renewable power capacity and aims to add an additional 60GW by 2030.

France plans spending of 7 billion euros within the next decade and Germany would invest 9 billion euros through 2040. The EU needs $150 billion in subsidies by 2030 to expand and $11 trillion of investment by 2050 (157).

Hydrogen $12 Trillion, One-in-a-Lifetime Market Opportunity

Goldman Sachs said: green hydrogen could be *"a nearly $12 trillion market by 2050".*

Gas pipelines and thermal power plans could get converted for green hydrogen. That investment led analysts to call green hydrogen a *"once-in-a-lifetime opportunity." The renewable energy market could grow tenfold in 30 years* due to the hydrogen demands. American company NextEra estimates that the potential for *"renewables and green hydrogen is 19 to 24 times the current renewable energy market"*. Many companies are stepping into the green hydrogen industry, pledging neutral emission by 2050 (158).

Natural Gas Hydrogen Injection

As part of the mega-project, *"a 5,000Megawatt"* renewable hydrogen export operation is being developed in Western Australia. The New Kalbarri project *"wants to mix the hydrogen directly with natural gas to lower its carbon footprint."* In 2020, the UK became the first country to implement *"grid injection of hydrogen,"* in Europe. A 20% volume blend allows customers to continue using their existing gas appliances (159).

Hydrogen And Transportation.

Hyundai Motor has just shipped 10 fuel cell electric trucks from South Korea to Switzerland. *"World's first fuel cell heavy-duty trucks,"* The Hyundai vehicles will change green hydrogen mobility in Europe. *"Hyundai will produce a total of 1,600 XCIENT trucks by 2025."*

The trucks are powered by a 190-kW hydrogen fuel cell system with dual 95-kW fuel cell stacks. Seven tanks with a combined storage of 32.09 kgs of hydrogen, allowing 400km in one refuel. Hyundai Motor is *"developing a long-distance traveling 1,000km on a single charge"*. Markets include North America and Europe. (160).

Train

Linde company will start the construction of the world's first hydrogen refueling station for passenger trains in Germany, in September 2020. 14 hydrogen-powered passenger trains will be fueled. The station will have *"a capacity of around 1,600 kg of green hydrogen per day."* This shows that hydrogen has an important role in decarbonization (161).

Airbus has unveiled three visual concepts for "zero-emission *airplanes"* to be powered by hydrogen. The *"zero"* initiative includes concepts for two conventional-looking aircraft:

-a turbofan jet engine, capacity 120-200 people over 2,000 nautical miles (3,700 km)

-turboprop capacity, 100 people for 1,000 nautical miles (nm), 1,650 km.

-a third proposal incorporates a *"blended wing body"* design (162).

Also, **ZeroAvia** company is creating hydrogen-powered commercial aircraft in the U.K, first practical, *"zero-emission airplane.* It is a short distance flight with just a few passengers, Unregulated, *"global aviation will produce 43 metric gigatons of carbon dioxide by 2050"*, equivalent to the global annual emission of 2019. Refueling these plans is an issue nowadays, the existing infrastructure is not compatible with fuel cell combustible. In the U.K. green hydrogen aviation developments will bring jobs to the economy, and reduce substantially emissions (163).

5.2.3. NUCLEAR ENERGY

NUCLEAR FISSION ENERGY

Nuclear Power Plants (NPP) based on the fission of atoms have been built after the Second World War. The catastrophic accidents from the USA, Ukraine, and Japan, stopped the building of new plants. In the last 60 to 70 years, the technology has advanced, so nowadays NPP can be built much safer, and more efficiently, being much smaller in size, reducing the time of construction, and most importantly the cost. Following are a few examples of new types of NPP, and SMR (Small Modular Reactor).

Small Modular Reactor

Leaders are looking at how to feed pollution-free heat to industries such as steel, cement, glass, and chemicals. 50% *of the world's energy goes into making heat",* and that produces *40% of the world's CO2 emissions",* based on IEA. Small modular reactors (SMR) are overlooked in the USA, Europe, and elsewhere. SMA can deliver a steady flow of heat and electricity. The heat decarbonizes the world's dirtiest industries. The biggest investors and industrial companies have gotten behind the technology with currently *"67 unique SMR technologies in various stages of development worldwide."* While modern reactors generate 1,500MW, SMRs have capacities of 300MW, and are designed to be mass-manufactured, which is a plus." Below is an example of SMR from NuScale company. SMR Will Change Energy, And Now It's Officially Safe. As mentioned above, NuScale has received a final safety evaluation report (FSER) for its modular reactor design, making it the first American modular design to reach this point. NuScale uses classic nuclear fission water reactor technology. NuScale says the reactor quenches itself in its pool, making it *"passively safe. Modularity and smallness are key aspects of NuScale and its peers".* The current design is for 50 MW per module. NuScale seeks to apply for a 60MW version. This is enough energy to power 48,000 homes. Fantastic (164).

UAE Connects First Arab Nuclear Power Plant to Grid

The United Arab Emirates has substantial oil and gas reserves, and now builds the first Arab country *"Barakah nuclear power plant."* On August 19, 2020, UAE connected its Barakah nuclear power plant to the national grid. Barakah, which means "blessing" in Arabic. Barakah was built by a consortium led by the Korea Electric Power Corporation at a cost of some $24.4 billion. The all-new Nuclear Power Plants have gained the attention of some politicians and investors, due to new technology, which is safer, reliable, and cost-effective (165).

Thorium Can Revive The Nuclear Energy Industry

For decades, the nuclear energy sector has been regarded as the black sheep of the alternative energy market thanks to a series of high-profile disasters such as Chernobyl (Ukraine), Fukushima (Japan), and Three Mile Island (The USA). Yet, nuclear alternative energy may survive through *"Substituting thorium for dangerous uranium in nuclear reactors."*

Thorium is producing less waste and more energy than uranium.

The United States Department of Energy (DOE), and 3 more partners started *"to develop a new thorium-based nuclear fuel called ANEEL".* ANEEL, which is short for *"Advanced Nuclear Energy for Enriched Life. A key benefit of ANEEL over uranium is that it can achieve a much higher fuel burn-up rate to the tune of 55,000 MWd/T (megawatt-day per ton of fuel) compared to 7,000 MWd/T for natural uranium fuel"* (almost 8 times more). For instance, India's Kaiga Unit-1 and Canada's Darlington PHWR Unit hold the *"world records for uninterrupted operations at 962 days and 963 days, respectively, (2.63 years). Thorium has a "much higher fuel burn-up that reduces plutonium waste by more than 80%".* ANEEL could soon become the fuel of choice for countries that operate CANDU (Canada Deuterium Uranium) and PHWR (Pressurized Heavy Water Reactor) reactors such as *"China, India, Argentina, Pakistan, South Korea, and Romania."* Overall, *"only about 50 of the world's existing 440 nuclear reactors can be powered using thorium"* (166).

Note: Capacity factor by energy sources: *"nuclear 93.5%, natural gas 56.8%, coal 47.5%, hydrogen 39.1%, wind 34.2%, and solar 24.5%"*

A New Vision of Nuclear Energy

Three factors require baseload power to support wind and solar energy. The rapid growth of EV cars, growing population, an increased rate of economic development, especially in developing countries. The answer is the current fourth-generation nuclear. A new 2GW nuclear plant can run for 60-80 years, during which time it can generate 0.9-1.2 trillion kilowatt-hours (kWh) of electricity without emitting pollutants. In the future, researchers anticipate lifespans exceeding 100 years. New modern nuclear plants cost $10.0 *billion for a 2GW plant".* That will drive the cost to competitive levels of $0.08-$0.1/kWh. *"The world needs roughly 1,500 new nuclear plants in a span of 50 years, that is 30 plants per year globally".* Currently in operation are 460 plants globally. All this tells us that humanity can cover in the next 8 to 10 years the baseload energy at a competitive cost and NO pollution using nuclear power (with a little luck maybe using fusion too). What about safety? The new generation nuclear plants are provided with a *"defense-in-depth,"* system. This consists of an independent system to check to prevent radiation release. After

1980, new technologies have made plants significantly safer. There are additional safety requirements implemented, which assure the safety of waste. We have access to large and affordable, low-cost, low-emission sources of energy. Through innovation in CCS and fusion technology, we can attenuate the climate crisis. The clear message is that we can not afford to abuse climate science, it is inhumane and is a crime against humanity. It was Cicero who in 80 BC first said, *"Such is the way of the world: no man attempts to commit a crime without the hope of profit"* (167).

NUCLEAR FUSION ENERGY

Fission nuclear reactors are based on splitting the atoms of nuclear fuel to produce lighter ones, and huge thermal energy, the fusion process is based on combining atoms of deuterium to obtain heavier atoms. Fusion reactions can generate unlimited energy. A multinational project (35 nations) to build a fusion reactor cleared a milestone, and is now 6 years away from *"First Plasma."* The International Thermonuclear Experimental Reactor (**INTER**), is built in southern France at the border with Swiss. The ITER project is the most *"colossal and ambitious scientific project in the entire history of humankind"*. It is an experiment aimed at reaching the highest stage in the evolution of nuclear energy. The project is in an advanced stage, as seen in the picture above (65% complete). I won't present technical details here, which can be found on the internet. Perhaps by the end of 2025, the project will be operational. *"It will take another 10 years until we reach full deuterium-tritium operations."* The project will contain the world's largest superconducting magnets.

UK Fusion Experiment

An experiment in Oxfordshire has been turned on for the first time. MAST (MegaAmp Spherical Tokamak) Upgrade could deliver clean, limitless energy for the grid. Unlike fission, MAST Upgrade will use an innovative spherical tokamak. Prof Ian Chapman said the switch-on was *"a momentous occasion. This moves the UK closer to building a fusion power plant."* The Culham Centre for Fusion Energy is home to both

MAST Upgrade and a fusion machine called JET (the Joint European Torus). JET uses a more conventional design. UKAEA's scientists now plan to test this new exhaust, Super-X diverter, at MAST Upgrade. This offers a *more compact fusion power plant* (169).

NASA And Nuclear Fusion Ambitions

"A NASA research project may offer a pathway to making nuclear fusion commercial." NASA tested *"lattice confinement,"* which could change production scale and take the costs down.

"NASA's lattice confinement method allows fusion-level kinetic energy to come together at room temperatures. The new method *"heats"* or accelerates deuterons sufficiently that when colliding with a neighboring deuteron, *"it causes D-D (deuterium-deuterium) fusion reactions"*. One of the competitors is *"dense plasma focus (PDF)"*, developed by LPPFusion. The cost of fusion is high. *"France's ITER has had an estimated cost of over $40 billion"*. Nuclear fusion could be the energy source opening up NASA's potential for greater space exploration along with partners such as SpaceX, Boeing, and Blue Origin (170).

Korean Artificial Sun.

The Korea Superconducting Tokamak Advanced Research(KSTAR), a *"superconducting fusion device, known as the Korean Artificial Sun,"* set the new world record as maintaining the high-temperature plasma for 20 seconds with an ion temperature over 100 million degrees. This is a core condition of nuclear fusion. The KSTAR began operating the device last August 2020 and conducting a total of 110 plasma experiments. In addition, the KSTAR Research Center conducts experiments and research on a variety of topics. The final goal of the KSTAR is to succeed in a continuous operation of 300 seconds with an ion temperature higher than 100 million degrees by 2025 (171).

5.2.4. GEOTHERMAL, WAVES, AND RNG ENERGY

Geothermal Energy

Is This The Cleanest Energy On Earth?

In 2018, the renewable energy advocacy group American Council on Renewable Energy (ACORE) *"launched an initiative to help secure $1 trillion in private sector investment in renewable energy."* Cumulative investment reached $125.1 billion through the first two years of the campaign. One renewable source hardly gets a passing mention: *"Geothermal energy."* Geothermal energy is ready to scale up and become a major player in clean energy. *Geothermal may hold the key to making 100% clean electricity available to everyone in the world".* It's an opportunity for the struggling oil and gas industry to use its capital and skills to work on something that won't pollute the planet (172).

Before we get to the technologies, though, let's take a quick look at geothermal energy itself.

What is Geothermal Energy?

The molten core of the Earth, about 4,000 miles down, is hot as the Sun's surface. The geothermal energy industry is called *"the sun beneath our feet."* The Earth heat has a flow rate of *"roughly 30 terawatts, almost double all human energy consumption",* and the process is expected to continue for billions of years. The ARPA-E project AltaRock Energy estimates that *"just 0.1% of the heat content of Earth could supply humanity's total energy needs for 2 million years."* As of the end of 2019, global installed geothermal electricity reached *"15.4 GW, with the US in the lead."* Geothermal energy is highly reliable since it's constant and available throughout the year, not depending on the weather. Geothermal power plants have average availability of 90% compared to ~75% for coal plants. More impressive *"it is one of the cleanest energy sources, and it is cheap."* The EIA says direct use applications and geothermal heat pumps have almost no negative effects on the environment. Consequently, *"Iceland's capital city, Reykjavik",* which heats 95% of its buildings using geothermal energy, is one of the cleanest cities on our planet. *"At US$ 0.04-0.14 per kWh, geothermal electricity has* *the lowest level cost of all U.S. generation sources.* Besides all otherworldly qualities, geothermal has some negativities.

-The shortage of high-temperature resources. It is known that the higher the differences between both ends (deep in the ground and surface) of the temperature, the more efficient the thermal power plants are. These high-temperature locations, closer to the surface of the planet, are rare.

-Another critical factor is that efficient thermal fields are very little in comparison with oil fields.

-And thermal electricity can not be easily transported, it has to be used locally. Due to these issues, getting thermal energy to the surface is difficult. The technology in this field has 4 categories of faulting. (14).

1). Conventional hydrothermal resources.

Hot springs, that water to get to the surface through the fissure of the edges of the tectonic plates, are rare. The few existing thermal fields are explored with conventional technology. In the USA such known fields are in California, Nevada, Hawaii, and Alaska. All these geotherms have their own local characteristics that make them difficult to standardize. That's why they are behind other renewable resources for so long. (14).

2). Enhanced geothermal systems (EGS)

The new technology involves drilling down into the rock, injecting the pressured water under the rock, and collecting the hot water through another well. This is called ESG, or geothermal that makes its reservoir. EGS are using fracking technology, without polluting the atmosphere or emitting GHG. For an average depth of 4.3 miles and a minimum water temperature obtained is 150°C. It is estimated that the total US geothermal resource is at least 5,157GW of electricity, around five times the nation's current installed capacity". If EGS technology is used, direct heat could provide the US with 15 million terawatt-hours-thermal (TWhth), equivalent to 15,000 million GW, 1GW = 1 billion Watts). *"Total US annual energy consumption is 1,754 TWhth for residential and commercial space heating,"* the US Department of Energy writes. *EGS may provide sufficient heat to every US home and commercial building for at least 8,500 years."*

There's enough heat down there to sustain civilization for generations. (14).

3). Super-Hot-Rock Geothermal

In the long run, EGS is a *"super-hot rock"* geothermal injection in deep, very hot rock. At much higher heat, the ESG Is incredibly high. When water exceeds 373°C and 220 bars of pressure, it becomes *"supercritical, a new phase that is neither liquid nor gas".*

What to know about supercritical water? *"First, it holds from 4 to 10 times more energy per unit mass".* And *"second, it is so hot that it "doubles the Carnot efficiency (principle in a combustion engine) of its conversion to electricity".* This way you can get *"more electricity out of that energy."* The difference between an ESG system running at 200C to double 400C is in energy efficiency from 5MW to 50MW. This supercritical water technology is in exploratory projects in Japan, Italy, Mexico.

4). Advanced geothermal systems (AGS)

AGS refers to a new generation of *"closed-loop"* systems, in which fluids circulate underground in sealed pipes and boreholes, picking up the heat through the conduction transfer principle, and carrying it to the surface. Here it is used for a mix of heat and electricity. Alberta (Canada) based company Eavor developed a system called an *"Eavor-Loop,"* like radiator design. Such ways permit the company to maximize the contact area of heating. With deep pipes to 1.5 miles, Eavor can get 150C water almost anywhere. Eavor can anytime complement variable wind and solar energy. Simply cutting off the flow of fluid, at the peak of solar and wind, the water stays in the system, heating up. When it is needed to turn it on the fluid is charging up, permitting the company to match almost any demand. Thermal energy has all advantages, it checks all the boxes (172).

Oil and gas to the rescue?

Geothermal may use oil and gas company infrastructure and qualified workers, such ways being a win-win solution. This is the best time to start or join a geothermal company. And there may be future billionaires companies. Oil and gas big polluters will buy geothermal companies and save part of their portfolio and assets. This way the big emitter will stay in power and produce the cleanest energy on the planet Earth. One part of the jobs lost in fossil fuel companies will be saved in the transition to geothermal energy. Geothermal energy may see more than $25 billion in investments in the next 5 years. Today's surface footprint of a geothermal plant is much lower in terms of square kilometers per MW of produced electricity. In Europe, several countries started to drill wells for geothermal plants. Germany also has several operational capacities, and will extend the drilling of 20 additional wells per year, *"using AGS vertical drills, depths as deep as "6000 meters."* The United States detains the most advanced studies, with a new climate policy, may subsidize more investments in geothermal energy (173).

WAVE ENERGY

Waves, As A Source Of Sustainable Energy

Ocean wave energy may be a dense alternative renewable source. Reza Alam, a wave energy researcher at the University of California at Berkeley says, *"solar"* farms receive *"0.2 to 0.3KW/sqm of incident energy"*. Wave energy is 10-15 times more efficient than solar and wind energy. Wave energy density can reach 30-40 KW/sqm on average and are always in motion 7/24. The availability of ocean waves is among its top advantages. *"Compare waves availability of 90% of the day* while solar *and wind energy are only available 20-30% of the day"*. Almost 2.4 billion people live within 60 miles of an oceanic coastline, where the wave farms are more effective. Wave energy systems require far less space. Wave energy converters like the Triton WEC (wave energy converter) from Oscilla Power are point absorber systems. The buoys are set in the water that moves with the waves, producing energy due to wave motion.

In 2015, the US Department of Energy held an 18-month long design-build-test competition called the Wave Energy Prize. Four companies including Oscilla Power exceeded the DoE's threshold. With its impressive availability and widespread application, wave energy's colossal opportunity is a future source of dense green energy. Investing In wave energy shall become a smart choice. *"Oscilla recently announced a financing campaign on July 1, 2020."* The company is looking to finance its commercial wave energy converter (WEC) project the Triton-C. The wave energy market will exceed $21 billion by 2027. *"The Triton-C is a 100 kW system"* for smaller applications, and for large applications, the *"Triton is a 1 MW system."* Wave energy has the possibility to become competitive with other renewable energies. Oscilla will operate in the U.S. and extend its operations to Europe and India (174).

RNG - Renewable Natural Gas Energy.

RNG - renewable gas natural, is an emerging alternative energy source that is produced *from the methane-rich animal, industrial, and food waste."* Thus, by increasing CO2 emissions but reducing methane emissions, *"RNG is net carbon-negative.* RNG does not need to build out new infrastructure to make it work. *"It is compatible with existing natural gas infrastructure".* One market to use RNG is the $800 billion global trucking market, where EV solutions are performing well, and hydrogen solutions are too expensive. Hyliion's electric powertrains are powered by RNG, a clean energy source, and cost-effective because current *"RNG prices hover around $1 per diesel gallon equivalent".* Hyliion's Hypertruck ERX, which is due to start deliveries in 2021/22. Compared with Tesla and Nikola, Hyliion's Hypertruck ERX is the best clean-energy long-haul truck technology in the market. Globally, about 1.7 million heavy-duty commercial vehicles (HCVs) are sold every year (175).

6

Impact of Global Warming, Actions

6.1. IMPACT (CONSEQUENCES) OF GLOBAL WARMING

The impact of climate change is spreading on each corner of the physical planet, and any angle of life on the Earth. From human life to biodiversity, ecosystems, ocean, ground, and air wildlife are disturbed, which will affect the safety and security of humanity and life on the planet. These all are interconnected on the planet like a neuron network in our brain, without considering borders, or regional limits of influences of governments or big industrial, and financial monopolies. The climate and weather phenomena don't have borderlines, same for birds, and fish. The impact is global, from China to America, and from Russia to Brazil. There is so much data about this issue, so it may be written in several books. The most important weather events and their consequences of global warming are presented in chapter 2, and paragraphs related to permafrost, peatland, and deforestation. Here additionally, I will present other impacts of the climate crisis on the planet, and human life.

6.1.1. EXTINCTION, BIODIVERSITY, AND ECOSYSTEM

Nature makes human development possible but our relentless demand for the earth's resources is accelerating extinction rates and devastating the world's ecosystems, and biodiversity. Humanity needs to integrate biodiversity considerations in global decision-making on any sector or challenge, whether it's water or agriculture, infrastructure, or business.

The following is a summary of the United Nations Report on The Global Assessment on Biodiversity and Ecosystem Services (IPBES, 29 April - 4 May 2017).).

The magic of seeing fireflies flickering long into the night is impressive. We draw energy and nutrients from nature. We find sources of food, medicine, livelihoods, and innovation in nature. Our well-being fundamentally depends on nature. Our efforts to conserve biodiversity and ecosystems must be underpinned by the best science that humanity can produce. The present generations have the responsibility to transfer to future generations a planet that is irreversibly destroyed and damaged by human activity. *"All our scientific knowledge is proving that we have solutions and so no more excuses: Earth is ours and we must live here as long as the Sun is giving us its blessing rays....* Humanity has to promote respect for life, and its diversity. We have to show ecological solidarity with other living species. Nature is deteriorating globally at unthinkable rates in human history, and the *"rate of species extinctions is accelerating"*. with unpredictable consequences on life around the world, warns the report from (IPBES). The summary of the report, which was approved at the 7th session of the IPBES Plenary, meeting (April – May 2017) in Paris, has raised everyone's eyebrows. The overwhelming evidence of the IPBES Global Assessment presents a unique picture.

"The entire planet ecosystem's health on which our interconnected species depend is catastrophically in peril. We are destroying the foundations of what is called our branches of making life possible." Nobody knows if it is not too late to change the existing status, but our duty is to do whatever it takes to save the climate crisis. *"Through transformative change, Mother Nature can be saved in order to be restored and used sustainably"*. Opposition to transformative change can be expected from those with material interests, (dollar man) but the public attitude has to stop it. The IPBES Global

Assessment Report on Biodiversity and Ecosystem Services is the most comprehensive ever completed (UN). It is the first Intergovernmental Report of its kind to present scientific evidence. Authored by 145 experts from 50 countries, with inputs from another 310 contributing authors. After systematically reviewing about 15,000 scientific and government sources, the Report also presents n indigenous and local knowledge, like groups from the Amazon forest or Arctic circle. Mother Nature's biodiversity contributes to humanity's *"most important life-supporting safety net"*. **The report finds that around 1 million animal and plant species are now threatened with extinction, many within decades, more than ever before in human history.** Since 1900 at least 20% of native species have disappeared. More than 40% of amphibian species, almost 33% of reef-forming corals, and more than 33% of all marine mammals are threatened. At least 680 vertebrate species had been extinct, and more than 9% of all mammals used for food and agriculture had become extinct by 2016, the Report says. *"Ecosystems, species, wild populations, local varieties, and breeds of domesticated plants and animals are shrinking, deteriorating, or vanishing. The essential, interconnected web of life on Earth is getting smaller and increasingly frayed,"* said Prof. Settele. *"This loss is a direct result of human activity and constitutes a direct threat to human well-being in all regions of the world."*

The authors of the Report have ranked the five direct changes in nature with global impacts. They are (1) changes in land and sea use; (2) direct exploitation of organisms; (3) climate change; (4) pollution, and (5) invasive alien species.

Notable findings of the Report include:

-Three-quarters of the land-based environment and about 66% of the marine environment have been significantly altered by human actions. On average these trends have been less severe or avoided in areas held or managed by Indigenous Peoples and Local Communities.

-More than a third of the world's land surface and nearly 75% of freshwater resources are now devoted to crop or livestock production.

-The value of agricultural crop production has increased by about 300% since 1970, raw timber harvest has risen by 45% and approximately

60 billion tons of renewable and nonrenewable resources are now extracted globally every year – having nearly doubled since 1980.

-Land degradation has reduced the productivity of 23% of the global land surface, up to US$577 billion in annual global crops are at risk from pollinator loss and 100-300 million people are at increased risk of floods and hurricanes because of loss of coastal habitats and protection.

-In 2015, 33% of marine fish stocks were being harvested at unsustainable levels; 60% were maximally sustainably fished, with just 7% harvested at levels lower than what can be sustainably fished.

-Urban areas have more than doubled since 1992.

-Plastic pollution has increased tenfold since 1980, 300-400 million tons of heavy metals, solvents, toxic sludge, and other wastes from industrial facilities are dumped annually into the world's waters, and fertilizers entering coastal ecosystems have produced more than 400 ocean *"dead zones"*, totaling more than 245,000 km2 (591-595), a combined area greater than that of the United Kingdom. Negative trends in nature will continue to 2050 and beyond in all of the policy scenarios explored in the Report, except those that include transformative change, due to the projected impacts of increasing land-use change, exploitation of organisms, and climate change, although with significant differences between regions.

-75%: terrestrial environment "severely altered" to date by human actions (marine environments 66%).

-47%: reduction in global indicators of ecosystem extent and condition against their estimated natural baselines, with many continuing to decline by at least 4% per decade

-28%: global land area held and/or managed by Indigenous Peoples, including >40% of formally protected areas and 37% of all remaining terrestrial areas with very low human intervention.

+/-60 billion: tons of renewable and non-renewable resources extracted globally each year, up nearly 100% since 1980.

15%: increase in global per capita consumption of materials since 1980.

-85%: of wetlands present in 1700 had been lost by 2000 – loss of wetlands is currently three times faster, in percentage terms than forest loss.

-8 million: total estimated number of animal and plant species on Earth (including 5.5 million insect species).

-Tens to hundreds of times: the extent to which the current rate of global species extinction is higher compared to average over the last 10 million years, and the rate is accelerating.

-Up to 1 million: species are threatened with extinction, many within decades.

-500,000 (+/-9%): share of the world's estimated 5.9 million terrestrial species with insufficient habitat for long-term survival without habitat restoration.

-40%: amphibian species threatened with extinction.

-Almost 33%: reef-forming corals, sharks, and shark relatives, and >33% of marine mammals threatened with extinction.

-25%: the average proportion of species threatened with extinction across terrestrial, freshwater, and marine vertebrates, invertebrates, and plant groups that have been studied in sufficient detail.

-At least 680: vertebrate species have been driven to extinction by human actions since the 16th century.

+/-10%: a tentative estimate of the proportion of insect species threatened with extinction.

-20%: a decline in average abundance of native species in most major terrestrial biomes, mostly since 1900.

+/-560 (+/-10%): domesticated breeds of mammals were extinct by 2016, with at least 1,000 more threatened.

-3.5%: the domesticated breed of birds will be extinct by 2016.

-70%: increase since 1970 in numbers of invasive alien species across 21 countries with detailed records.

-30%: reduction in global terrestrial habitat integrity caused by habitat loss and deterioration.

-47%: the proportion of terrestrial flightless mammals and 23% of threatened birds whose distributions may have been negatively impacted by climate change already.

-6: species of ungulate (hoofed mammals) would likely be extinct or surviving only in captivity today without conservation measures.

The report also presents a wide range of action needed to be taken by governments (policies, climate laws, regulation, etc.), financial institutions, all big fossil fuel companies, industry, and sectors that pollute and emit GHG in the atmosphere (UN Report).

Extinction. *Sir David Attenborough makes a stark warning about species extinction.*

Note: I am sure that Sir David Attenborough will be more than delighted to read my notes in this book about his colossal personal experience with species extinction.

Britain's favorite naturalist is not celebrating the large diversity of life on Earth but to issue everyone with a stark warning." *We are facing a crisis,"* he warns at the start, *"and one that has consequences for us all."* What follows is a shocking reckoning of the damage our species has wrought on the natural world. There are the stunning images of animals and plants you would expect from an Attenborough production, but also horrific scenes of destruction.

In one sequence monkeys leap from trees into a river to escape a huge fire.

In another, a koala limps across a road in its vain search for shelter as flames consume the forest around it. Of the estimated eight million species on Earth, *"a million are now threatened with extinction, one expert warns".* Since 1970, vertebrate animals, birds, mammals, reptiles, fish, and amphibians, have declined by 60%. The world's last two northern white rhinos feature in the film. These great beasts used to be found in their thousands in Central Africa but have been pushed to the brink of extinction by habitat loss and hunting.

"Many people think of extinction being this imaginary tale told by conservationists," says James Mwenda, *"but I have lived it, I know what it is."* James strokes and pets the giant animals but it becomes clear they represent the last of their kind when he tells us that Najin and Fatu are mother and daughter. What a sad story. Species have always come and gone, that's how evolution works. But, says Sir David, the rate of extinction has been rising dramatically, 100 times the natural evolutionary rate, and is accelerating. *"Throughout my life, I've encountered some of the world's most remarkable species of animals,"* says Sir David, in one of the most moving sequences in the film. *"Only now do I realize just how lucky I've been - many of these*

wonders seem set to disappear forever." Sir David is at pains to explain that this isn't just about losing the magnificent creatures he has featured in the hundreds of programs he has made in his six decades as a natural history filmmaker. The loss of pollinating insects could threaten the food crops we depend on. Trees and other plants regulate water flow and produce the oxygen we breathe. Meanwhile, the seas are being emptied of fish. There is now about 5% of trawl-caught fish left in oceans and rivers, compared with before the 1900 year. *"Our job is to report the reality the evidence presents,"* But the program does not leave the audience feeling that all is lost. Sir David makes clear there is still cause for hope. *"His aim is not to try and drag the audience into the depths of despair,"* but to take people on a journey that makes them realize what is driving these issues so we can also solve them."* One of the most celebrated moments in all the films Sir David has made features the moment he met a band of gorillas in the mountains on the border between the Democratic Republic of Congo and Rwanda. A young gorilla called Poppy tries to take off his shoes as he speaks to the camera. *"It was an experience that stayed with me,"* says Sir David, *"but it was tinged with sadness, as I thought I might be seeing some of the last of their kind."* We learn that the Rwandan government has worked with local people to protect the animal and that the gorillas are thriving. There were 250 when Sir David visited in the 1970s, now there are more than 1000. It shows, says Sir David, what we can achieve when we put our minds to it. *"I may not be here to see it, but if we make the right decisions at this critical moment, we can safeguard our planet's ecosystems, its extraordinary biodiversity, and all its inhabitants."* His final line: *"What happens next",* says Sir David, *"is up to every one of us."* (176). A few additional examples of animal extinction will take place in the next 2 to 3 decades if humanity does not act.

Whale. The killing of endangered North Atlantic whales is far faster than previously thought. This will lead to extinction if we do not act. Of the remaining 356 whales, only about 70 breeding females survive. Experts fear females could disappear in the next 10 to 20 years. Again humanity must act. *"To see them turning up dead or even worse, entangled in ropes where it takes a year to slowly die, is just gut-wrenching."* There has to be a wake-up of the social and political consciousness of humanity. In the last few years, the Canadian government has taken steps to reduce the decline of whales, including limiting the speed of large ships and closing commercial fishing areas. The experts warn that more must, and can be done.

Sharks and Rays are disappearing at an *"alarming"* rate. The number of sharks found in the open oceans has *"plunged by 71% over half a century"*, especially due to overfishing. 3/4 of the species studied are now threatened with extinction. And again and again, more actions are required to secure a normal-natural future for these *"extraordinary, irreplaceable animals"*.

Overfishing is the main reason for the decline of more than 70% of the shark population in the last ½ century. Out of 31 species studied, *"24 are now threatened with extinction,"* and three shark species have declined so deeply they are now classified as critically endangered, according to the International Union for Conservation of Nature (IUCN). Of the 1,200 or so species of sharks and rays in the world, 31 are oceanic. Humanity needs to protect these biggest predator species in the oceans. The political will to save these species comes from pressure from the planet's citizens. *Action across the globe will prevent catastrophic consequences and assure the future for these extraordinary, irreplaceable animals."* (177).

Millions of Seabirds Killed in the Pacific

A *"blob"* of unusually warm seawater in the Pacific Ocean was responsible for the deaths of about 1 million seabirds in 2015 and 2016, a study released on January 17, 2020, said.

The birds, a species known as the common murre, died of starvation due to the warm water, which severely disrupted the food supply. *"The magnitude and scale of this failure have no precedent,"* said study lead author John Piatt, a research biologist in the U.S. Geological Survey's Alaska Science Center. Ocean warming is a red flag that climate change impacts on seabird's disappearing, and the safety of ocean ecosystems. Ocean heat wave created a blob of warm water in 2013, which significantly affected marine life. Other species experienced mass die-offs after blob warmth appeared, including tufted puffins, sea lions, and baleen whales.

The size and temperature of the blob were astonishing: *'2F(1.12C) to 7F(3.9C) degrees above average temperature, and about "1,000 miles across and 300 feet deep"* (178).

The list of the examples of extinction of animals, birds, plants is long. I will stop here, the reader has made his/her mind what is tragic, devastating, and inadmissible for the extinction of part of life on Earth.

6.1.2. SEA LEVEL, ARCTIC ICE, GLACIERS, AND CORAL REEFS

Ice Melting

Losing the Greenland ice sheet would be catastrophic to humanity. a 20 feet (6.1 meters) rise in ocean level will be the consequence of ice melting. That would change the shape of coastlines around the planet. Almost *"40% of the global population lives within 60 miles (96 km) of the coast, and 600 million people live within 30 feet (9.13 meters) off sea level"*. As consequences will be the relocation of communities, climate refugees, and costly infrastructure including homes will be abandoned. New research from the *"journal The Cryosphere reports,"* shows the rates of melting ice are accelerating and compatible with the Intergovernmental Panel on Climate Change's (IPCC). Between 1994 to 2017, approximately 28 trillion tonnes of ice (both poles) melted (equivalent to "*1.2 trillion tonnes of ice per year*)" and 2/3 *due to the warming of the atmosphere*, and 1/3 was due to rising sea temperatures. That ice loss caused global sea levels to rise by around 3.5 centimeters (1.4 inches).

Iceberg A68 And A74

Back in December 2020, reports warned of a 1,620-square-mile iceberg, (4147. sq. km.), called A84, which broke off from the Antarctic peninsula, was on course to collide with South Georgia Island. In doing so, it would crush coral sponges and plankton on the seafloor. Also, this will trigger cutting off seals and penguins from their normal feeding. As it turns out, "*warmer waters shattered*" the iceberg missing South Georgia Island. As a result, "*the penguins and seals will be spared from the collision.*" Through the melting process, an influx of cold freshwater would get into the ocean, killing off phytoplankton, a disaster for the food chain. The krill that feed on phytoplankton could starve, which will start the depleting *populations of fish, seals, penguins, and whales*" More *icebergs break off from the Antarctic will cause the rising global temperatures*" (169). Latest, iceberg A74 broke off from the Antarctic too in February 2021.

Coastal Cities Flooding

Hundreds of millions of people are at risk of losing their homes, as cities around the globe will be submerged under rising sea levels over the next 30+ years.

The **Nature Communications** journal shows that some 340 million people will be living on land that falls below sea levels by 2050. Major cities like Bangladesh, Shanghai, Tianjin, Hong Kong, Jakarta, and Thailand, risk being uninhabitable. These cities need to take as soon as possible action to avoid the "*economic and humanitarian catastrophe*". The same situation is facing many cities in the coastal area of the USA.

Arctic Ice

A study suggests that the Arctic *"may be essentially ice-free during summer within 15 years."*

The study used statistical models to predict the future amount of Arctic ice that will be melted by the year 2034. Sea ice affects Arctic communities and wildlife such as *"polar bears, and walruses"*. Also, sea ice regulates the planet's temperature by influencing the circulation of the atmosphere and ocean currents. *"Polar bears may disappear"* if arctic sea ice keeps shrinking. When the Arctic has less than 1 million square kilometers of sea ice it will be the first ice-free Arctic. Climate change causes the Arctic *to warm twice as fast as the rest of the planet"*. Arctic air temperatures were 3.4C, (6.12F) degrees above average in 2019 and were the second-warmest since records began in 1900 (179).

Canada's Arctic Ice Shelf Collapses

The last fully intact ice shelf in the Canadian Arctic has collapsed, losing an astonishing percent of more than 40% in just two days at the end of July 2020. The shelf's area shrank by about 80 sqkm, compared to the island of Manhattan in New York which covers roughly 60 sqkm.

"This was the largest remaining intact ice shelf, and it's disappearing," Siberia fire, Canadian Arctic, which has been 5 degrees Celsius above the 30-year average, and less ice less reflected sun's rays contributed to the melting. These incredible changes in the Arctic in the last 40 years have not been encountered since 3 million years ago, Earth's climate may return to the Pliocene era. The consequences: *"Florida and California's Central Valley would be underwater, and it would be too hot to grow corn and wheat in the Midwest and Great Plains".*

The lowest sea ice cover takes place each year in September. In 2020, it measures 1.44 million square miles (3.74 million square kilometers), 50% of the area covered in 1980. This is the lowest value in the last 42 years, *"since satellites began taking measurements"* (180).

Glaciers

Melting glaciers in **Alaska** could unleash a *"mega-tsunami"*, a colossal high wall of water with unknown destructive power. When the glacier melts, they destabilize mountain slopes, fall into the water and trigger a mega-tsunami that could be unleashed as early as next year, 2020.

Collapsing mountain slopes in the US state in 2015 had led to waves up to 633 feet (208 meters) in height. *"The mountain slope above the toe of Barry Glacier in Barry Arm, 60 miles east of Anchorage, has the potential to fail and generate a tsunami in 2021,"* or later, a group of researchers warned the Alaska authorities. or later. *The tsunami will be devastating and highly destructive, through landsliding into the fiord below."* (181).

Alpine Glacier.

Alps Swiss glaciers have continued to shrink at an alarming rate this year 2020, while snow accumulation has recorded the lowest level, a study showed Friday, Oct. 16, 2020. The glaciers in the Alps are losing 2% of mass per year, which is a lot of it accumulates from year to year.

The most concerning finding are that *"snow accumulation on the mighty Aletsch, the largest glacier in the Alps, hit its lowest level on record."*

"It is cold up there, but there was very little snow that was retained, and this is, of course, a bad sign for the biggest glacier in the Alps." The glacier *"has already seen its tongue recede by about one kilometer since the turn of the century"*. Aletsch out of more than 4,000 glaciers is a vast, ancient reserve of ice, providing seasonal water to millions. *"If greenhouse gas emissions are left unchecked, by 2100 "around 95% of those glaciers will disappear.* Even if humanity does not do anything to slow down global warming, 67% of the Alpine glaciers will likely be lost (182).

The Himalayas are sacred mountains. Their name in Sanskrit means *"abode of snow"*. But glaciers are a political issue in central Asia. *"Glacier-fed rivers provide water to over a billion people for food production and hydropower"*. Any disruption in the normal cycle of snowing during winter and melting during summer will be catastrophic for those billion people. India and Nepal are relying on glacier meltwater to sub-press spring droughts. India, which opposes climate change, has presented evidence that glaciers in northern India and Pakistan are stable. In the area of the Himalayas, the political situation is very unstable, deterring access for field observation. The melting of glacier areas was misleading and concealed. When satellite Earth observation technology opened, it started to get the correct data about glaciers melting. The scale of glacier change across the Himalayas could be seen for the first time. The new data obtained by satellite revealed that nearly all Himalayan glaciers were shrinking at a similar rate. *"In the last 20 years the rate at which glacier ice is lost has doubled, and is similar to the rate of ice loss globally."* In this way, we learned that *"high mountains are warming twice as fast as the rest of the planet"*. It is estimated that 1/3 and 1/2 of the Himalayas' ice glaciers will be lost by

2100. The afferent countries of Himalayas glaciers are now conscious of the loss of ice. They have to take action too as the entire society (183).

The Latest Himalaya Glacier Broke.

Sunday, February 7, 2021, part of a mountain glacier broke, sending a massive flood of water and debris slamming into two dams and damaging several homes. At least 200 people were killed and an unknown number were missing. The sliding wall of water happened when a portion of the Nanda Devi glacier broke off in the Tapovan area. A video of a steep hillside shows *"a wall of water surging into one of the dams and breaking it into pieces with little resistance"*. *On the Alaknanda river,* the Rishiganga hydropower plant was destroyed, and the Dhauliganga hydropower was damaged (183).

Himalaya Tsunami. In June 2013, record monsoon rains in Uttarakhand caused devastating floods that claimed close to 6,000 lives. That disaster was called the *"Himalayan tsunami."* The torrents of water sent mud and rocks crashing down, burying homes, sweeping away buildings, roads, and bridges. Environmental experts called for a halt to big hydroelectric projects in the state. *"The government should no longer ignore warnings from experts and stop building hydropower projects and extensive highway networks in this fragile ecosystem"* (183a))

Underwater Ice

A new study trying to find out how much ice melts beneath the water surface, using incorporated sonar and time-last photography. The study duration was almost 3 years. In some places, the *"glacier melted 100 times faster than previous models predicted"*. For three years, they used the methodology to survey the LeConte Glacier in Alaska. In some places," *the researchers recorded the glacier melting 100 times faster than previous"*, theoretical models predicted. This tells us that existing models are inaccurate to study tidewater glaciers. That leads to a more devastating global sea-level rise. Previous estimates have to be adjusted, not only for some glaciers but all tidewater glaciers.

Coral Reefs

After recovering from the 2020 record fire, Australia's *"Great Barrier Reef experienced its third mass coral bleaching event in the past five years"*. High water temperatures and marine heatwaves, makes coral bleaching a normal occurrence. Coral reefs are the most vibrant, and sensitive ecosystems on the planet. The year 2020 was one of the hottest years on the planet, causing more coral bleaching in the world. Higher water temperatures and pesticide exposure may be affecting the development of baby reef fish. The transition from baby reef fish to reefs is encountered with external unhealthy conditions for their development. These negative effects, warmer water, and pesticides are endangering entire coral reef communities (184).

Australia Great Barrier Reef. By 2100, we may not find coral reefs on the planet due to global warming, according to research released in February 2020.

Renee Setter, a biogeographer at the University of Hawaii at Mānoa, said: By 2100 ocean life is looking quite grim. In the study, researchers simulated ocean environments, based on projections of sea surface temperature, ocean acidification, wave energy, pollution, and fishing practices. The findings show that by 2100, few to zero suitable habitats for corals are likely to remain. Only a few sites that will support reefs by 2100 are small parts of Baja California in Mexico and the Red Sea. In 2016 and 2017, *"about 50% of coral in Australia's Great Barrier Reef died after record heat triggered mass bleaching"*.

Coral reefs are an integral part of ocean ecosystems, supporting thousands of fish and other marine species. Besides climate change, coral reefs are threatened by illegal fishing, coastal development projects, and pollution. (185).

Saving Coral Reefs. Scientists in Australia found a way to help coral reefs fight the effects of bleaching by making them more heat-resistant. Warmer water makes corals expel tiny algae which live inside them. This turns the corals white and effectively starves them. In response, researchers have developed a lab-grown strain of microalgae that is more tolerant to heat. When injected back into the coral, the algae can handle warmer water better.

The next step is to further test the algal strains across a range of coral species. (186).

6.1.3. HEALTH, FOOD, AND WATER

HEALTH AND CLIMATE CHANGE

Climate Change Is The Biggest Health Threat This Century.

More medical schools are training doctors to recognize and treat the effects of climate change.

We know that the last five years (2016-2020) have been the hottest on record, nearly 1.7F (1.2C) warmer than the 20th-century average. The students examine, for instance, heat-related mortality, as well as the healthy *"upside"* of adapting to the effects of global warming. *"Reducing one's reliance on cars and eating smarter benefits the human body, and Mother Earth alike"*, Patz argues. There are two dozen medical schools leading which are training physicians, nurses, pharmacists, and other practitioners to respond to climate change-induced diseases. The spread of *"toxins, asthma cases, cardiovascular disease, and Lyme disease"* is on the rise. That is *"the biggest global health threat of the 21st century"*.

Air pollution is responsible for about 8.7 (updated) **million deaths a year** worldwide, according to the World Health Organization (WHO). Reducing the burning of fossil fuels could avoid *"2.5 million premature deaths"* each year by 2050. Physicians also need better training in screening for vector-borne illnesses (infectious pathogens, thick ticks, or mosquitoes). Environmental changes also affect respiratory disease, cardiovascular disease, and cerebrovascular mortality, or stroke, said Harvard's Jha. There's also mental health to consider, such as the decision to have children. A new study dated January 9, 2021, shows that air pollution caused by burning fossil fuels was responsible for 8.7m deaths globally in 2018. The enormous death toll is higher than previous estimates and surprised even the study's authors. Without fossil fuel emissions, the average *"life expectancy of the world's population would increase by more than a year, while global economic and health costs would fall by about $2.9trillion."* A major report by the Lancet in 2019, for example, found 4.2 million annual deaths from air pollution. *We don't appreciate that air pollution is an invisible killer"* (187).

Rising Heat Waves Death

According to EcoWatch, a new study from the Climate Impact Lab found that by the end of 2100, *"extreme heat could kill roughly as many people as all infectious diseases combined, including HIV, malaria and yellow fever"*. Older people die due to indirect heat effects.

The economic costs of all the deaths will easily *"cost the world 3.2% of its total economic output"* without controlled emission. The researchers calculated that *"each ton of carbon dioxide emitted will cost $36.60 in damage."* One conclusion of the report states that *"poorer countries will bear the brunt of the suffering from extreme heat"* (188).

Wildfire Smoke, And Health Issues

Excessive heat and fire in the western US, during the 2020 worst fire season, increased hospital admissions *"for asthma to the university's healthcare system rise by 10% and cerebrovascular incidents such as strokes jump by 23%"*. Also will climb the number of heart attacks, kidney problems, and even mental health issues. Australian fire from 2019 to 2020 found a bleaker outcome. Researchers from the University of Tasmania *"identified 417 extra deaths due to smoky air, and reported 3,100 more hospital admissions, and 1,300 extra emergency room visits for asthma."* A pandemic crisis lapping on top of the climate crisis, it's difficult to endure and manage. The results of fire are very fine particles, known, PM2.5, which causes the biggest concerns. The fire's smoke puts you at a greater risk for infectious diseases. *"It requires a healthy immune system to fight infections like Covid-19.* Since 2006, the heatwave in *"West Coast has killed approx. 600 people, put 1,200 more in the hospital and sent 16,000 to emergency rooms,"* according to Southwest Climate AdaptationHistoric (189).

Post-Covid Syndrome

The Covid-19 is creating an aftermath health crisis, called *"post-Covid syndrome"*.

Post-Covid-19 analysis by top American scientific bodies has found that many *"patients who recovered from the coronavirus infection are experiencing deadly health issues like heart damage, stroke, neurological problems, lung damage or pulmonary fibrosis, chronic fatigue syndrome and multi-system inflammatory syndrome in children."* A study by the Journal of American Medical Association conducted in Frankfurt, Germany found *"78% of the recovered Covid-19 patients developed heart-related problems".* The Same situation has been discovered by UK studies in 69 countries, which found *"heart abnormality in 55% patients while 15% patients showed severe abnormalities",* who need further treatment, like oxygen support from time to time for the rest of their lives (190).

Children, And Young Adults Health vs. Climate Change

An U.N.-backed report said in February 2020, *"warning climate posed an urgent threat to the health and future of every child and adolescent".* Countries with high carbon emissions, including the USA and Australia, are at high risk of health-related climate change. The heating up of the atmosphere and pollution expose every child to uncertainty and health issues. Countries with a long period of time with high pollution like Idia, China, Lagos, Nigeria, are exposing their children to earlier chronic diseases. This is a problem for the future generation of their population (191).

Heat Waves And Mental Health.

Heatwaves can also harm our mental health in hidden but surprisingly severe ways. Heatwaves tend to make our blood boil. Historic studies found that hotter regions tend to have higher violent crime rates than cooler regions. This link between heat and other factors that affect violent crime rates, like poverty, unemployment, age distribution, and culture, accentuates the crisis. Temperature increases are associated with increased suicide rates. Above 18°C, each 1°C increase in temperature is associated with a 3.8% increase in the incidence of suicide. For instance, during the 1995 heatwave in the UK, suicide increased by a staggering 46.9%. Countries with more than 45°C heat, like India and Pakistan, have already suffered an immense loss of life and livelihood, and the emotional trauma of the aftermath. Millions of people with pre-existing mental health conditions are exposed to greater health risks during heatwaves The issue will accelerate with more intense heat waves coming (192). It became clear that every year to come, the health issue for humanity will be a burden to society, more costly, more death and suffering on top of the other issues like food and water insecurity, housing problems due to flooding, hurricanes, and wildfire destruction. All these repeating events will create a continuous state of mental stress, affecting everyone, poor or wealthy, young or adult, white, brown or black. The sickness doesn't make a difference or has limits or borders.

Food And Water Security

Global warming will affect the food and water supply to people in any corner of the planet.

A quarter of the world's population across 17 countries are living in regions of extremely high water stress. Qatar, Israel, and Lebanon were ranked as the most water-stressed countries in the world. A global water crisis will require better information, planning, and water management. Populations are growing, and the economy requires more water supply, but the supply is diminished by climate change, water waste, and pollution. The water crisis will trigger a security crisis, which can lead to regional friction, even conflicts, migration, and other disrupting events. The global research organization compared the water available to the amount withdrawn for homes, industries, irrigation, and livestock.

"In the 17 countries facing extremely high water stress, agriculture, industry, and municipalities were found to be using up to 80% of available surface and groundwater."

When demand is higher than supply, small quantities of reduction in supply (caused by climate change), may produce dire consequences. India, the Middle East, and North Africa are the most affected countries. It is possible that other countries in different regions of the planet may become severely affected. In the US *"New Mexico and California,"* are facing a very low supply of water, and the issue is increasing day by day, as drought continues to persist. *"New Mexico state's score is on par with the United Arab Emirates and Eritrea."* Drought in California is in the 10th consecutive year, which creates misery in slow motion (193).

There are many issues and conflicts between parties, like community and industry rights of use of natural water resources. On the other hand, many communities can not use polluted water by industry, from here evolving all kinds of legal issues. But the big problem in most cases is the depletion of sources of water due to global warming. For instance, Himalaya glaciers melting affects more than *"1.2 billion people in Southeast Asia."*

Food Crisis

"The world must diversify its food production and consumption or face damaging supply disruptions that could result in suffering and social unrest, even nations security instability," scientists have warned. *"A small disruption in food supply leads to huge price increases",* said Per Pharo of the Food and Land Use Coalition (FOLU). *"Four different crops provide 60% of our calories, wheat, rice, maize, and potatoes".* The present damage done by food supply to health, economic development, and environment costs the world *"$12tn (£9.6tn) a year."* For instance, as a measure to protect the forests in Costa Rica, the government has reversed deforestation by eliminating cattle subsidies and introducing payments to farmers. The area of the forest has risen from ¼ to ½ from the country area in 1983. Overall the cost

estimated was $350 billion annually and created a business of 4.5 trillion, a 15-fold return. The reforms have to be implemented anywhere where it is possible, like the saying: "Where is a wish, is away." (194).

Perfect Storm - Farming, And Marine Fisheries

Nearly *"90% (7.2 billion) of the world's population ... are projected to be exposed to losses of food production by 2100"*. In contrast, less than 3% (0.2 billion) of the population live in regions that would experience productivity gains. Regions like Latin America, Central, and Southern Africa, and Southeast Asia, would disproportionately face losses in both farming and marine fishery sectors. *"These areas are generally highly dependent (that leads to cost increase), on agriculture and fisheries for employment, food security, or revenue"* (195).

Food Crisis, And Pandemic

An estimated 265 million people could go hungry in 2020, nearly double the 2019 figures, according to the World Food Program (WFP)'s projection in April 2020. As millions around the world are losing their jobs or seeing their incomes cut, it's increasingly difficult for them to afford food. Lockdown measures are making it harder to transport and commercialize food. Refugees and people in conflict zones like *"Yemen, Syria, and Burkina Faso and those already living hand-to-mouth are particularly vulnerable."* The pandemic is also affecting countries that mainly rely on tourism, services, or manufacturing jobs are likely to be hard-hit. Several countries, including Russia, restricted exports of key food commodities. In these situations, countries are using artificial methods like food price increases. So I might say, looking at your neighbor dying by innaniton is not the best idea (196).

One Country Could Disrupt Global Food Supply

The USA is trading wheat with 174 countries. Catastrophic crop failures caused by extreme weather (4 years drought) in just one country, the USA, could disrupt global food supplies. This is how climate change threatens global stability. A study found the United States would deplete nearly all its wheat reserves after four years of drought and affect all countries where America is exporting its crop. In other wheat-producing countries like Russia and France, encountering droughts. Global warming could increase the frequency and intensity of droughts. Five years of recurring droughts have destroyed maize and bean harvests in Central America in 2019. In 2020, 2 hurricanes devastated Central America, destroying everything in their way. Both these weather events resulted in thousands of people trying to immigrate north to the USA (197).

Iowa's Corn Crop Lost to Derecho

The occurrence of a rare land wind storm that slammed Iowa in August 2020 has increased by more than 50%. a new report shows. The U.S. Department of Agriculture said that the number of "crop acres" *that Iowa is unable to harvest has grown to 850,000 acres (343,983 hectares).*

The storm, known as a derecho, generated winds of up to 140 mp/h (225 km/h) that flattened crops. *"The total cost of the storm is evaluated at $22 billion"* (198).

American Diet: 'We Would Need Another Planet'

"If everyone on the planet consumed an American diet, some 138% of the planet's surface would be needed to produce the food, scientists found". In the last 50 years, the amount of
"meat eaten globally has quadrupled, exceeding 320 million tonnes per year", the researchers said. Ph.D. student Hanna Pamula was looking at how meat consumption impacts land use, water and air pollution, and the carbon dioxide equivalent. Eating just one serving of chicken, pork, beef, lamb, or fish a week can have a devastating impact on your carbon footprint, the researchers said. *"Five servings of meat a week wastes: 14,363 liters of water, the amount of water 13 people drink in a year."* Pamula said: *"I'm not trying to convince everyone to go vegan."* Even a small reduction in meat consumption can make a noticeable difference (199).

Additional Data For This Paragraph.

300%: the increase in food crop production since 1970.

23%: land areas that have seen a reduction in productivity due to land degradation.

-75%: global food crop types that rely on animal pollination.

US$235 to US$577 billion: the annual value of global crop output at risk due to pollinator loss 5.6 gigatons: annual CO_2 emissions sequestered in marine and terrestrial ecosystems – equivalent to 60% of global fossil-fuel emission.

+/-11%: undernourished world population.

-100 million hectares of agricultural expansion in the tropics from 1980 to 2000, mainly cattle ranching in Latin America (+/-42 million ha), and plantations in Southeast Asia (+/-7.5 million ha, of which 80% is oil palm), half of it at the expense of intact forests.

-3%: the increase in the land transformation to agriculture between 1992 and 2015, mostly at the expense of forests.

-33%: world's land surface (and +/-75% of freshwater resources) devoted to crop or livestock production.

-12%: the world's ice-free land used for crop production.

25%: the world's ice-free land used for grazing (+/-70% of drylands).

+/-25%: greenhouse gas emissions caused by land clearing, crop production, and fertilization, with animal-based food contributing 75% to that figure.

+/6-30%: global crop production and global food supply provided by small landholdings (2 ha), using +/-25% of agricultural land, usually maintaining rich agrobiodiversity (UN, WFP Report).

6.1.4. MIGRATION, AND POPULATION

Climate Change and Migration.

By the middle of this century (2050), experts estimate that *climate change is likely to displace between 150 and 300 million people.* No individual country nor the global community is completely prepared to support a whole new class of *"climate migrants."* The U.S. government is spending

US$48 million to relocate residents of Isle de Jean Charles, Louisiana because their land is sinking. *"Climate migration is already happening".* Every year desertification in Mexico's drylands forces 700,000 people to relocate. Experts agree that a prolonged drought may have catalyzed Syria's civil war and resulting migration. *"Between 2008 and 2015, an average of 26.4 million people per year were displaced by climate- or weather-related disasters (UN)".* And the science of global warming shows that these trends are likely to get worse. More than 410 U.S. cities and others around the globe, like Amsterdam, Hamburg, Lisbon, and Mumbai are vulnerable. Rising temperatures in west Asia make it inhospitable to human life. And these examples can go on and on. Today the UN has not universally acknowledged the existence of climate migrants. We don't know how to define them. United Nations refugee law, *"climate migrants are not legally considered refugees".* Climate migrants don't have any protections that are accorded to refugees, or agreement to help them. A refugee is defined as someone *"unable or unwilling to return to their country of origin owing to a well-founded fear of being persecuted for reasons of race, religion, nationality, membership of a particular social group, or political opinion."* International shells consider *"modifying international law to provide legal status to environmental refugees."* Also *"the world's largest carbon polluters could contribute to a fund that would pay for refugee care and resettlement"* for those displaced. All issues of climate migrants need to be discussed in Glasgow IPCC in Nov. 2021.

For instance, by midcentury, climate change will create more refugees than World War II, which displaced some 60 million across Europe (200).

1 Billion People Displacement By 2050?

Climate change disasters will displace *"more than 1 billion people by 2050."* Compiled by the Institute for Economics and Peace (IEP), uses data from the United Nations, to *"assess eight ecological threats and predict which countries and regions are most at risk."* World's population will reach 10 billion by 2050, the research shows as many as *"1.2 billion people living in vulnerable areas of sub-Saharan Africa, Central Asia, and the Middle*

East" will be forced to migrate by 2050. The threats can be grouped into 2 categories:

1). food insecurity, water scarcity, and population growth,
2). natural disasters including floods, droughts, cyclones, rising sea levels, and rising heat.

For example, the world now has 60% less freshwater available than it did 50 years ago, while *"food demand will reach 50% more in the next 30 years"*. Those factors, combined with natural disasters mean even stable states' stability is at risk by 2050.

Latest on Climate Refugee (March 17, 2021).
About *"10.3 million people were displaced by climate change-induced events"* such as flooding and droughts *"in the last six months (up to March 2021)"*, mostly in Asia. *About 2.3 million others were displaced by conflict in the same period."* These figures are only for six months from September 2020 to February 2021. *"In some instances, people barely have time to recover from one disaster and they're slammed with another disaster."* McKinsey & Co consulting firm has said that Asia *"stands out as being more exposed to physical climate risks than other parts of the world".* Globally, *"17.2 million people were displaced in 2018 and 24.9 million in 2019.*
I do believe that these three reports tell everybody enough about the risk humanity faces regarding the overlapping of more than 2 crises (pandemic and economic) on top of the ongoing climate crisis. It is a scary picture of life on the planet by the middle of this century (201).

Population Growth vs. Climate Change

Some people believe that the world is too crowded and keeping a lid on population growth could be an answer to tackling climate change. Many experts disagree.
The argument is that if there were fewer people on Earth, greenhouse gases would be reduced and climate change could be averted. But experts say *"population control isn't the panacea,"* some think it might be. *"It is a very complicated, multifaceted relationship. Population issues are the main factor*

in how society will develop, but are not the critical cause in climate change".
If it takes 123 years for the population to go from 1 billion to 2 billion
people, for the last billion needed only one decade. Of course, population
growth is an important issue in climate change, but slowing population
growth alone *"won't solve climate change".*

Global birth rates are declining. The United Nations projected that
the global population would reach 11.2 billion by 2100, and in June 2020,
the prediction dropped to 10.9 billion. That's because fertility rates are
dropping for most of the world. Another big factor for the population
decline is urbanization. In the 1960s, about 33% of the population lived
in a city; now it's 54% and by 2050, that will be 68%. In the cities, it is
more expensive and difficult to raise children. When people move to the
city, they tend to have fewer children. We don't need to eliminate half of
the Earth's population to end starvation. We need to educate people in
undeveloped countries, *"that women do not have 7 kids, and see them die of
starvation".* One of the concerns of climate change is the effect of global
warming on the population. Today's generation hopes that their kids will
have a better life. This should be the norm of life (6). A very interesting
thing to take into consideration about dying. The lifespan of people is
increasing, especially in developed countries. In China, the average life
expectancy is now 76.5, and by the mid-2030s, the prediction is 80 years.
*"The key to achieving slower population growth is through educating, family
planning and reproductive health services."* Millions of women don't have
control of their fertility, and access to the information and services that will
allow them to do that" (202). Education is a decisive factor in the control
of human behavior for climate change, as well as fertility and control of
the population (202).

6.1.5. SECURITY OF THE PLANET

Is The Planet Under Security Threatened?

British naturalist David Attenborough has warned on Tuesday,
February 23, 2021, that *"climate change is the biggest security threat that
modern humans have ever faced",* telling the U.N. Security Council: *"I
don't envy you the responsibility that this places on all of you."* Attenborough,

94, the world's most influential wildlife broadcaster, addressed a virtual meeting of the 15-member council on climate-related risks to international peace and security, chaired by British Prime Minister Boris Johnson. *"If we continue on our current path, we will face the collapse of everything that gives us our security: food production, access to freshwater, habitable ambient temperature, and ocean food chains,"* Attenborough said. *"And if the natural world can no longer support the most basic of our needs, then much of the rest of civilization will quickly break down,"* he added. The United Nations will stage a climate summit in November 2021 in Glasgow, Scotland. The summit will decide "to be or not to be " life on the unique planet called EARTH. Glasgow will decide the fate of humanity. If Glasgow fails, humanity will fail its fate and not deserve the luxury of life. Another type of humans needs to replace us, those with respect for Mother Nurture and conscious of their responsibility for belonging to such a colossal and magnificent place in the Universe. *"It is our last, best hope to get on track and to get this right,"* U.S. climate envoy John Kerry told the council. He continued: *"We bury our heads in the sand at our peril. It is time to start treating the climate crisis like the urgent security threat that it is. This is the challenge of all of our generations."* The November summit is the time for the countries to commit to their existence. Those who believe that this is a place for denial of climate change, are the enemy of themself. Russia and China are two big emitters of GHG and try to find any hidden reasons not to comply with the Paris agreement. China's ambassador to the UN described climate change as a development issue. How can the UN hold these countries accountable for their facts and attitude? So far as I know there are not only two countries that hide their emission. U.N. Secretary-General Antonio Guterres told the council. *"This is a credibility test of their commitment to people and the planet"*. Climate change means security in the world. If climate change fails, security fails. Does that mean security fails in the world? Where there are nuclear weapons to destroy the planet more than ten times. Here stays the importance of the Glasgow summit (203 UN Security Council).

Ecocide an International Crime

The Paris agreement is failing. Yet *"there is new hope for preserving a livable planet: the growing global campaign to criminalize ecocide can address the root causes of the climate crisis and safeguard our planet"*. *The climate agreement five years ago did nothing to slow down the climate crisis.* The last study shows the increased temperature tipping point will be reached in less than 5 years. Carbon in the atmosphere reached 417 parts per million (ppm), the highest in the last 3 million years. *"Even if we magically flipped a switch to a fully green economy tomorrow, there is still enough carbon in the atmosphere to continue warming the planet for decades"*. In other words, if humanity does not take *"drastic measures"*, *life on the planet is under dramatic threat of existence. Besides physical actions, humanity must take action inside its own social policies and rules.* In that direction, establishing **ecocide** as a crime is as mandatory as existentialism. Global jurisdiction must enact laws against ecocide. Ecocide has to be not only on words and paper, it has to be criminalized by-laws, to set it at the same level of genocide. Earth ecological destruction is the elimination of indigenous communities. For instance, the Yanomami in Brazil and native Alaskan villages are decimated by poisoning mercury or erosion of their soil by drilling on their lands for mining. The ecocide action will require that the International be able to act against such eco-disasters like deforesting the Amazon and Congo basins, drilling in the Arctic and the Niger Delta, or illegal plantation of palm oil in south-east Asia. Public organizations *are calling on global leaders to introduce ecocide at the International Criminal Court (ICC)"*. French President Emmanuel Macron vowed to introduce this ecocide law in his country. Finland, Belgium and Spain follow in French shoes. *"And Pope Francis was ahead of the game in November 2019 when he called for ecocide to become an international crime against peace"*. The Stop Ecocide Foundation is working with international lawyers to implement the ecocide amendment to the ICC's Rome Statute. This will criminalize the reckless leaders from acting against the Paris agreement measure. Lately (May 2021) a Duch court ruled that oil and gas company Royal Dutch Shell has to reduce its GHG (CO_2) emission by 45% from 2019 levels by the year 2030. Shell has proposed a reduction of emission to 20%, less than half what the court ruled out. This is the first unprecedented legal action in the world of big emitters, and a sign of consent for the other big emitters (204).

Indeed, the meaning of ecocide is fully encapsulated by its etymology. It comes from the Greek *Oikos* (home) and the Latin *fall* (to kill). Ecocide is literally *"killing our home (planet)"* (204A International Criminal Court UN).

Climate Change Could Seriously Weaken the U.S. Military.

Energy shortages in the U.S. curtailing training and even the ability to deploy troops overseas.

A new report shows a grim picture of the Earth's future, including U.S. security. The report, commissioned by the Army *"to explore areas likely affected by climate"*. Some effects: at home, disruptions in the power grid caused by wildfires could cut military bases off from electricity, the use of fossil fuels by the Pentagon, stockpiles of fuel for nuclear weapons. Abroad: the Military will be sent to restore order in a world pushed into disorder. Climate change could force the migration of millions of people. Also, diminishing natural resources like freshwater may trigger conflict. Rising temperatures will accelerate the spread of tropical diseases. The report urges the Pentagon to begin preparing for climate change. The Department of Defense *"is worried about the future of energy and the instability of a planet in a climate crisis"*.

In an article titled *"The U.S. military is quietly going green,"* business news site Quartz reported that *"U.S. defense leaders recognize the threat of climate change, even if their commander-in-chief doesn't."* The article goes on to say that *"*American military strategists have long viewed *climate change as a "threat multiplier." Climate change is a security threat primarily because it multiplies and complicates existing security risks,"* says independent military news site Stars and Stripes. Military installations are vulnerable to flooding, droughts, and fire. Also, *"the magnitude of DoD's investment in energy RDT&E reflects the importance of energy to the military mission"*. The Department of Defense believes that climate change is a real, relevant, and present threat (205).

The Pentagon (January 2021). Over the past decade, *"the U.S. military* *and intelligence officials have developed a broad agreement about the security threats that climate change presents."* Biden's focus on climate change has cheered international partners and environmental advocates but upset Big Oi. The cynicism of Big Oil companies is not giving up, and it is presented very well in the book *"The New Climate War, The Fight to Take Back Our Planet"* by Michael Mann. It is a national security issue, and we must treat it as such, *Austin said.* Biden has called climate change an *"existential threat"* and promised to spend *"$2 trillion to expand clean energy and build resilient facilities over the next four years".* The Pentagon's more than $700 billion budget may offer an opportunity for the Biden administration to help scale up industries suitable with climate change. The grid blackouts proved the country's energy infrastructure failure, as the January 2021 polar vortex storm hit, especially in Texas, putting several million people at a security risk. The inability of the U.S. power grid to withstand the stresses caused by extreme weather events shows that the nation needs a *"massive investment plan to upgrade energy infrastructure"* and not only. The actual grid failures show the *existing* (largely fossil-based) system cannot handle these climate conditions. The blackouts in Texas in February 2021, and in California in January 2021, show that the current grid in the United States needs an overhaul update. Current policy in these states and others, can't stop the weather from wreaking havoc and putting people's lives at risk (206).

Doomsday Clock

The Bulletin of the Atomic Scientists has updated its Doomsday Clock to 100 seconds to midnight, the same time the group set it to the last year 2020. The Doomsday Clock isn't updated on a set time frame, but rather, as events dictate. You can thank the pandemic, climate change, and the threat of nuclear war for this update. Former Manhattan Project scientists created the Doomsday Clock in 1947 (6). Life as we know it is on the brink of disaster, according to the Bulletin of the Atomic Scientists, a nonprofit organization made up of scientists and global security experts. On January 27, 2021 morning, the group published a new statement deriding the global response to the COVID-19 pandemic and expressing concern about *"nuclear weapons and climate change".* The setting is the closest we've come to a symbolic apocalypse since the first tests of the hydrogen bomb in 1953. *"Still, the pandemic serves as a historic wake-up call, a vivid illustration that national governments and international organizations are unprepared to manage nuclear weapons and climate change, which currently pose existential threats to humanity".* The ways humanity has invented to destroy itself have multiplied since 1947 and in 2007, the *Doomsday Clock began to consider climate change as a dire challenge to humanity."* The Doomsday Clock is one of the rarest things available to scientists: an easily recognizable icon that can grab a passerby with no scientific background. In short, it's exactly what Rabinowitch and Goldsmith wanted (6).

I presented here more data from the United States, where data is available, but sure the G20 countries, and more developing countries are facing the same issues of their security.

6.2. ACTION TO SLOW DOWN CLIMATE CHANGE

6.2.1. GENERAL NOTES.

The Earth is, beyond a reasonable doubt, careening toward climate catastrophe in a man-made crisis propelled by the technological advancements that enabled us to use fossil fuels and then exacerbated by a dangerous level of political and economic myopia. The new president of the USA, Joe Biden will order that climate change be part of policy decisions across the federal government. To succeed in global warming, Congress needs to involve itself here to both stabilize how the nation approaches energy policy and to craft laws that will force the U.S. into a faster and more robust transition from fossil fuels to renewables. We won't get prescriptive about specifics (though here's one: Congress ought to revoke its 2017 approval of drilling in the Arctic National Wildlife Refuge), but some broad contours are clear. I believe that Congress does not need to wait for a *"catastrophic weather disaster with millions of casualties"* to start implementing climate laws. Same for the rest of the governments around the planet.

We need the governments to intervene in the energy system, from production to consumption, through laws and regulations. We need to get more aggressive and creative in designing green new buildings, retrofitting old structures, and converting homes from heating with natural gas or oil to renewable electricity. Yes, we can hear the sputtering. *"This is going to be insanely expensive but it's unavoidable."* Remember, it was a technological leap, harnessing the energy of burned coal to run industrial machinery, that led us to this point, and we are entirely capable of fresh technological breakthroughs to lead us to a different future. *"It's numbing, and depressing, to contemplate how little the U.S. and the world have done to combat global warming in the five years since the Paris agreement".* The past brought us here. The present will determine our future.

Bidenomics

The most important event in the world at the end of 2020, was the election of former vice president **Joe Biden**, as the New President of the United States of America. Hours after his inauguration on January 20, 2021, President Joe Biden *"signed an executive order to rejoin the Paris Agreement,"* ending the United States' exit from the global pact. It was the third executive order he signed in the Oval Office. In separate executive orders, Biden is expected to lay the groundwork for the part that comes next: reversing the Trump administration's deregulatory legacy and setting ambitious *"new goals for decarbonizing the U.S. economy."* The executive actions include orders to pull permits for the Keystone XL oil pipeline, reestablish the Interagency Working Group on the Social Cost of Greenhouse Gases, and direct all agencies, and government departments to immediately review all rules and regulations guidelines. He nominates climate envoy John Kerry, to take concrete actions to point to as he embarks on his new climate-focused diplomatic mission. The administration's *"trade agenda will hinge significantly on its relationship with China".* Brokering a *"cooperative peace between the U.S. and China will be key to any global decarbonization plan".* China's manufacturing capacity for clean energy products is unmatched and could be expanded. *"If we want to transition the global economy away from fossil fuels as fast as we need it to, we need China to be doing a lot of the heavy lifting because of how strong its new-energy production capacity is. The U.S. would be able to move a global Green New Deal diplomatically and would be able to line up financing for it in ways that would be complementary to what China brings to that project."* (207).

Today, American technology is in another race for world leadership. This time the race is for climate solutions. Across the globe, nations are ready, willing, and able to invest in the proven products and technologies needed to meet their Paris Agreement commitments. *"The country that can claim that leadership will see its economy dominate the future"*

Yet again, America is starting the race well behind the leader. China's technological climate solutions are far in advance of our own in four key areas:

Solar energy: China has an installed solar power capacity of over 240 gigawatts, twice that of the U.S. They continue adding more, twice as fast as we are.

Ultra-high-voltage transmission: China has created a million-volt transmission line serving power to Beijing from over 2.000 miles away. The U.S. has nothing like it.

Electric transport: China had 4.6 million electric vehicles on the road by December 2020 compared to just 1.74 million in the U.S. They also have 23,000 miles of high-speed trains to our 343-mile Acela train that runs between Boston and D.C. at much slower speeds.

Hydrogen Power: China produces twice the manufactured hydrogen as the U.S. Both countries are researching carbon-free hydrogen for truck, ship, and jet fuel plus hydrogen-based energy storage to play key roles in the climate solution (White House).

A Price on Carbon: Put a price on carbon emissions equal to the economic harm they do to the climate. That allows the market to prioritize investments in carbon-free alternatives. Passing a carbon fee and dividend policy, such as the Energy Innovation Act, would provide the economic base for our moon shot. (208).

An Advanced Transmission System: Build out an ultra-high-voltage, long-distance energy transmission infrastructure to efficiently move wind and solar power (from Idaho and the West) to wherever the nation might need it.

Green Innovation Investments: Make massive R&D investments in energy storage, carbon-free steel and cement, low-emission agriculture, and green hydrogen fuel cells.

More importantly, since China (No. 1) and the U.S. (No. 2) are the world's biggest carbon emitters, *"the competition might save the planet from its looming climate disaster."*

Another approach the Biden administration would be to commit to funding and deploying technologies and policies to remove carbon from the atmosphere. Carbon removal may be nature-based like planting new forests and adopting new soil farming techniques to technological solutions such as Direct Air Capture machines, DAC, or CCS (Carbon Capture and Sequestration), that suck CO_2 from the atmosphere (209).

Direct air capture systems are pretty expensive, to build an efficient one to annihilate emissions from a big oil company reaches *"approximative $1.2 trillion, (capture approx. 12 billion tons of CO_2 = 0.012 Gigatons at $100/ton)"*. If new technology can lower the price under $100/ton of CO_2, the cost will be lower too. While there's debate over how many Gigatons of carbon need to be removed from the atmosphere, like 10 Gigatons of carbon per year, many experts agree on the need to deploy technology such as direct air capture. Other experts believe that we need to take from the air a minimum of 30 Gigatons per year so that the balance of existing carbon in the atmosphere does not increase. In 2019 it emitted 39 Gigatons of GHG.

There's an earnest scientific debate over how much more carbon dioxide the atmosphere can absorb before locking in 1.5C of warming. At the end of 2017, U.N. scientists projected that a carbon budget of just 420 Gigatons remained in the atmosphere (420 billion tons), and humanity has roughly spent that at a rate of *"42 Gigatons per year."* (The budget of carbon in the atmosphere has values which vary a lot. Some scientists go with 1.2 trillion metric tons, others are going for 900 billion tons, or even 300 billion. In my conclusion I will take the UN report of 420 billion tons) (210).

All US Government Departments to Take on Climate Change

U.S. President Joe Biden has promised an *"all-of-government"* approach to fight climate change. The strategy marks a departure from past Democratic administrations that tended to run their climate strategies mainly through the Environmental Protection Agency (EPA). The interagency approach could help Biden make headway on global warming without having to rely on sweeping climate legislation that would be hard to pass in a divided Congress.

"Every agency is a climate agency now," Here are how some federal agencies could likely play a role in the administration's climate agenda:

-The EPA, as the nation's top environmental regulator, will be the obvious choice for Biden as his administration seeks to beef up climate protections and enforce regulations. The EPA could also rebuild rules and scientific processes that had been systematically dismantled by the previous administration.

-Interior Department. In charge of a fifth of the country's surface, the Interior Department will carry out Biden's campaign pledge that the United States will conserve nearly one-third of all land and water by 2030. Biden has promised to halt new oil and gas drilling on federal lands and waters, a policy that will need to be led by the Interior.

-Department of Energy. The Energy Department will lead the charge on clean energy research and development programs, including advanced nuclear, fusion, battery, and biofuel technology. Biden has said he intends to direct massive public investment into cleantech to ensure the country is able to eventually hit net-zero emissions by 2050.

-Defense Department. The DOD is the single largest buyer of fossil fuels in the fed's and can use its enormous procurement powers *"to purchase clean and resilient energy technologies".*

-Treasury Department. The Treasury Department could use the Dodd-Frank Wall Street reform act to require banks and other institutions to weigh climate risk in their investments, something that could trigger a big swing in private investment flows toward clean energy. *"I think the Treasury Department is going to be a pretty big player on climate,"* the dollar man prevails.

-Agriculture Department. Agriculture accounts for 9% of U.S. emissions but U.S. forest land, much of which is overseen by the Agriculture Department, sequesters up to 15% of annual fossil fuel emissions. The department could pay farmers, ranchers, and forest landowners to reduce and offset emissions through a credit program. It could also use crop insurance policies to encourage *"climate-smart agriculture"*, according to policy experts advising Biden.

-Education Department. The Education Department could direct federal dollars toward the funding of specialized teachers and programs to raise awareness of climate change. *"There could also be lots of opportunities for new dollars for K-12 and higher and technical education.*

-**Justice Department.** The Justice Department will have to defend the new administration's climate policies but could also prioritize federal civil cases with a climate focus. Biden has also said in his climate plan that he will form an environmental and climate justice division within the DOJ to prioritize climate change and environmental justice (208).

Climate Adaption.

"An organization that promotes efforts to adapt the environment to cope with the effects of climate change is calling on governments and financiers around the globe to include funding for adaptation projects in their COVID-19 recovery spending." The appeal was published on January 22, 2021, in a report issued by the Netherlands-based Global Center on Adaptation. *"As governments begin spending trillions of dollars to recover from the pandemic, the world has a once-in-a-lifetime opportunity to build a more resilient, climate-smart future by integrating climate adaptation into their response and recovery plans,"* the center said in its report. A group of more than 3,000 scientists from 130 countries also released a statement before the summit linking investment in the environment with pandemic recovery plans. *"The twin threats of COVID-19 and climate change are, above all, caused by human actions. We must do everything in our power to ensure our response to both is coordinated and becomes a watershed moment for investment in a more sustainable world,"* scientists wrote. Former U.N. Secretary-General Ban Ki-moon said that the two-day event hosted by the Netherlands *"couldn't be more timely."* Our recovery from the COVID-19 pandemic will take a year or so, *"but climate crisis recovery will take centuries"* of hard work in the energy transition, consequences of climate distraction, human behavior for a better society (209).

6.2.2. CARBON PRICE, AND CARBON TAXES

Social Cost of Carbon

"The social cost of carbon might be the most important number on climate change"

Economists are urging the U.S. government to adopt a higher number for the social cost of carbon emissions. The social cost of carbon is one that helps decide *"how much we're willing to invest to slow down global warming, and how much we value the future"*.

Two prominent economists Nicholas Stern and Joseph Stiglitz published a paper making the case that the U.S. needs to reassess how it calculates the social cost of carbon.

"The social cost of carbon reflects the ultimate estimated dollar price to society for every new metric ton of carbon dioxide emitted". Stern and Stiglitz, putting more value on the wellbeing of future generations, *"suggest a social cost should be $50–$100 per metric ton range"*.

What they're saying: *"It is vital to get the number right, and by right, we mean higher than it has been in the past,"* The social cost of carbon "represents the economic benefit that will come from reducing carbon emissions". The number is also the best representation of what today's generation is doing for future generations in terms that future generations will benefit from our action on climate. Stern and Stiglitz, both citing ethical reasons to adopt a lower discount rate which will generate a higher social cost of carbon. Experts are saying that the *"estimate could run as high as $125 per metric ton"* and go higher each other 6 months. Fossil fuel industries and manufacturing sectors are opposing such rates which fall on their shoulders.

The bottom line is: *"Where we set the social cost of carbon tells us how we put a value price on a future-focused threat, and how much we value the generations to come"* (211).

Global Climate Change Cost in 2020

Cost of disasters exacerbated by climate change produced $210 billion of losses in 2020 according to a report by the reinsurance company Munich

Re. The worldwide monetary losses in 2020 were up 26.5% compared to 2019's cost of $166 billion. In the year 2020, 8,200 lives have been lost due to global warming, higher than the previous year. Reinsurers, which provide financial protection to insurance companies, saw their share prices drop over the course of 2020. Shares of Munich Re, Berkshire Hathaway, Swiss Re, and AIG dropped 1.5%, 2.7%, 14.3%, and 21%, respectively.

U.S. Faces Year of Record Damage

Natural disasters in the U.S. accounted for $95 billion of overall losses in 2020, compared to $51 billion in 2019, an increase of 86.3%. Of the 10 costliest global natural disasters last year, six occurred within the U.S. For instance hurricane, Laura, hitting Louisiana in August 2020, costing $13 billion, of which only $10 billion was insured. North Atlantic was hyperactive, with 13 out of 30 storms reaching hurricane status. In total, losses from the hurricanes in North America, in 2020 came to $43 billion. The Midwest storms cost $40 billion, while losses from the wildfires in California, Colorado, and Oregon amounted to approximately $16 billion. The Vortex in Texas in February 2021 is estimated to cost between 195 and 295 billion dollars.

Floods in China produced losses of $17 billion, but only around 2% of that was insured. Total losses in Asia were $67 billion. Europe's natural disasters were: $12 billion of overall losses and $3.6 billion of insured losses (212).

Global Carbon Price. Can you put a price on pollution? Some of the world's biggest economies are trying to do just that. China, Japan, and South Korea have all followed the European Union's pledging to cut emissions to *"net-zero, where they release only as much as they remove from the air". "Each country will have to come up with its proposal to reach net-zero, but carbon pricing is going to be the most difficult part of it,"* said Wendy Hughes, Carbon Markets and Innovation Manager at the World Bank. The principle is simple: a carbon price establishes how much companies need to pay for their emissions. The higher the price, the better it is to reduce pollution and invest in green energy. A carbon tax by governments make companies pay their degree of polluting, *or through an emissions-trading* system (ETS)". An ETS sets a maximum cap on the quantities of emissions that a sector can produce. This is similar to *"carbon permits"* companies are buying for each tonne of CO2 they emit. *"The ETS has proven its efficiency,"* said Mr. Timmermans, head of EU climate policy. Price balancing act. The trick is to get the carbon price right. If it's too low, no dent in pollution, too high risks of efficiency of the industry. The carbon price, currently at around 27 euros per tonne of CO2 (this sounds like a joke, too low), needs to hit higher levels to determine companies to lower their emission. And the carbon price should be *"around 90 euros ($108 at Feb. 2021 rate) a tonne is a reasonable expectation by 2030."* China plans to adopt its own national ETS, possibly as early as 2021. *"Establishing a single, global price on carbon remains a wanted wish."* So far this is like putting them together in one enclosure of the population of a zoo. The EU agrees to introduce border-tax, for polluting goods and protecting their industry (213).

How Carbon Prices Will Transform Industry.

With freak weather, crop failure, and forced migration, climate change is imposing on humanity ever-higher costs. Governments, businesses, and investors are trying to find a solution on how to impose a fair sustainable cost on the emitters. The world has to get together and establish a carbon price, this is the most difficult monetary issue the nations must agree on. Carbon pricing is dictating decarbonization, by incorporating the cost of greenhouse gas emissions into the price of goods and services. One giant oil corporation is suggesting a carbon price of $100/tonne of GHG by 2030. This is a good proposal before 2025. In June 2021, bankers, investors, and fossil fuels companies have to pass the test of the carbon price, but all in Asia, Europe, and other regions are blaming the cost is much too high. *"A $100 price"* might bankrupt 1000 largest companies globally equivalent to $2.1 trillion (according to calculations for Lex by consultancy Planetrics, part of Vivid Economics). The issue raises more questions to industry, banks, investors, and governments. The policymaker should pay attention to the induced changes in behavior, and putting a monetary value on future environmental damage. To suppress opposition to carbon price may be done through recycling revenues from carbon taxes or issue of pollution permits in benefits or tax cuts. This approach of "carbon dividends" has attracted bipartisan support in the US. The decarbonizing financial impact will not spread equally. Industries with energy-intensive consumption will pay more than others. The cost of higher emissions sectors such as steel, utilities, cement, and power should be much higher.

Globally will be *"winners and losers within large economies*, the Planetrics analysis found. Among the energy sector, coal will be the big loser with 90% of its value, and renewable energy will be gaining more than double its value. The energy sector will be the hardest hit with an impact of 40%. But, *a business-as-usual forecast, which will lead to passing climate tipping points which will be catastrophic for the Earth."* Generally, a global carbon price to cover all sectors of economies is very complicated and requires diversification per sector and government subsidiary. Major U.S. financial trade groups on February 18, 2021, agreed to some type of carbon pricing in the United States. *"For markets to function you've got to have a carbon price,"* said Tim Adams, president of the Institute ofInternational Finance. Its board includes leaders of banks, insurers, and asset managers worldwide, including JPMorgan Chase & Co, Citigroup Inc, and State Street Corp. A policy document will be released showing that carbon pricing can *"spur development of climate-related financial products, promote more transparent pricing of climate-related financial risks, and can inform and help scale key initiatives like voluntary carbon markets."* Other backers include the American Bankers Association, and the Investment Company Institute, representing top asset managers.

The Biden administration has revived the idea of taxing carbon emissions, and Congress is crafting legislation. In America, which has two-party lines, it is difficult to get an easy consensus on the carbon tax. President Joe Biden wants to include an *"enforcement mechanism"* for reducing emissions. Any carbon tax bill triggers spending required by progressives to support the bill. The bill has to take into consideration the effects of low-income and communities of color. Those environmental justice concerns have gained traction nationally in recent years. The democrats favour a carbon tax as detrimental to the economy. They are backed by a report issued by the National Academies of Sciences that endorsed policies in pursuit of reaching net-zero carbon emissions by 2050. *"We need to work through and find a solution that works best and can pass into law,"* Whitehouse said (213).

"Unfortunately, the climate crisis does not have a blue or red color."

China Trading Market.

On February 1ˢᵗ, 2021, China flipped the switch on a nationwide carbon trading market, in what could be one of the most significant steps taken to reduce greenhouse gas emissions. China is the first big country to implement a national carbon emissions trading market. The carbon emission trading system includes China's power industry and approx. 2,000 energy generation facilities which *"represents 30% of the nation's total emissions"* according to ChinaDialogue.

Still, China is putting more weight on the economic benefits rather than the environmental costs of much of its industrial growth. Prices are expected to start at 41 yuan (US$6.3) per ton of HGH and rise to 66 yuan per ton in 2025. The real carbon price is estimated between $40 and $80 by 2020, and somewhere in the $50 to $100 range by 2030. *"The hardest part of carbon pricing is often getting it started. The moment that the Chinese government decides to increase ambition with the national ETS, it can. The mechanism is now in place, and it can be ramped up if the momentum and political will provided by President Xi's climate ambition continues. In the coming years, this could see an absolute and decreasing cap, more sectors covered, more transparent data provision, and more effective cross-government coordination. This is especially so with energy and industrial regulators who will need to see the ETS not as a threat to their turf, but as a measure with significant co-benefits for their carbon price policy objectives"* (204).

As presented, the carbon price is a very complicated issue, which depends on several factors: economical, social, technological, political, and so on. At the same time, emission control at a global level is the most possible solution to slow down the climate crisis. There are many aspects of carbon pricing, and data for different regions or countries, different types of industry, and people behavior, but the most acute factors have been shown. If these 3 big emitters, China, the U.S.A., and the EU, agreed on April 22, 2021, a virtual summit to celebrate Earth day. Unfortunately at this summit, only words have been changed, no actions taken. The importance of the issue determined me to pay special attention to this aspect of the climate crisis and provide detailed data (214).

6.2.3. DECARBONIZATION, NET ZERO

General Notes

The worldwide effort to prevent Earth from becoming an unlivable hothouse is in the grips of *"net-zero"* fever. *"More than 110 countries have committed to becoming carbon neutral by mid-century"* according to the United Nations. The European Union has taken the vow, as has US President Joe Biden. *"More than 65% of global CO2 emissions now fall under such pledges"*, according to a UN estimate. The London-based Energy & Climate Intelligence Unit calculates the aggregate GDP of nations, cities, and states with *2050 net-zero targets is "$46 trillion, well over half of global GDP. I firmly believe that 2021 can be a new kind of leap year, the year of a quantum leap towards carbon neutrality,"* UN chief Antonio Guterres said in Dec. 2020 in New York. *"Every country, city, financial institution, and company should adopt plans for transitioning to net zero emissions by 2050. In many cases, net-zero pledges are an improvement, but in others, the 'net' provision is a black box that can conceal all sorts of problems,"* Duncan McLaren, a professor at Lancaster University's Environment Centre, told AFP. Earth's surface has already warmed 1.3C on average, making extreme weather more deadly, and *"new research shows that a return to 2019 levels of carbon pollution would likely push the world past the 1.5C milestone around 2030."* The new studies put this value of 1.5C before 2025 (author note May 30, 2021). The devil is in the details, said Kelly Levin, a senior associate with the World Resources Institute's (WRI) global climate program. If you plant trees today and they will mature in 25 years, it is a different bird than capturing GHG directly from the air. *"There are several keys to evaluating the worth of carbon-neutral promises"*, Levin and other experts said.

-The first is whether they *"apply to all greenhouse gases or just carbon dioxide"*. CO2 is responsible for more than 75% of global warming. Also, concentrations of methane, mostly from natural gas leaks and agriculture, are rising and may overturn the balance of neutrality.

- A second red flag is the *"lack of intermediate hard targets before 2050"*, said Teresa Anderson, climate policy coordinator for ActionAid International. Scientists are categorical about the *"need for deep, near-term reductions in GHG"*. It is a huge difference to make a reduction for a time frame exponentially with the time curb holding water or not. The UN's IPCC has said that emissions must drop 45% by 2030, and then 100% by 2050. The UK was announcing a 68% cut in carbon emissions by 2030, encouraging other leaders to follow suit. The European Union could boost its 2030 pledge to 55%.

-A third most important measure is how much of percentage will be fulfilled with short-term emissions cuts, and how much will come from so-called *"negative emissions technologies".*

"You cannot get to net-zero without some carbon dioxide removal," or CDR said Oliver Geden, a researcher at the German Institute for International and Security Affairs, and an IPCC lead author, which is true and very true mathematically speaking. The process can be tricky. Planting trees, which will mature after 2050, to absorb HGH efficiently, does not work and needs an area twice as large as India. Nowadays technology ccs and DAC are at an infancy stage. What is remaining is to substitute primary energy with renewable energy, use energy efficiently and cut pollution through innovation in all sectors of energy production and consumption. **Finally**, *"one must read the fine print of net-zero pledges to see exactly what is, or is not, included".* As I will present in the next paragraph, direct carbon removal from the atmosphere must be a predominant process in carbon-neutral achievement. In other words, global warming's impact will be as bad as they are now unless we remove GHG to reduce the emission of gases (215).

United Nation

"Humans waging 'suicidal war' on nature" - UN chief Antonio Guterres.

"Our planet is broken," the Secretary-General of the United Nations, Antonio Guterres, has warned. Humanity is waging what he describes as a *"suicidal war on the natural world."*

"Nature always strikes back, and is doing so with gathering force and fury," he told a BBC special event on the environment. Mr. Guterres wants to put tackling climate change at the heart of the UN's global mission and *"declared a "State of Emergency for the entire planet".*

In a speech entitled State of the Planet, he announced that its *"central objective"* the next year 2021, will be to *"build a global coalition around the need to reduce emissions to net-zero".*

The objective, said the UN secretary-general, will be to *"cut global emissions by 45% by 2030"* compared with 2010 levels. Here's what Mr. Guterres demanded the nations of the world do:

-Put a price on carbon

-Phase-out fossil fuel finance and end fossil fuel subsidies

-Shift the tax burden from income to carbon, and from taxpayers to polluters

-Integrate the goal of carbon neutrality (a similar concept to net-zero) into all economic and fiscal policies and decisions

-Help those around the world who are already facing the dire impacts of climate change

"The science is clear: unless the world cuts fossil fuel production by 7% every year between now and 2030, things will get worse. Much worse." The 7% cutting is not enough if we take numbers in the calculation. Society is pumping up 40 Gigatons of GHG annually. Reduced with this percentage next year the emission will be 37.2 Gigatonnes, and so on. The planet these days can not absorb even half of the emitted pollution. It is clear that the amount of GHG in the atmosphere will grow from year to year and not decrease. Today existing GHG in the atmosphere is evaluated to be approx. 420 Gigatons. If in the next 10 years we pump an average of 20 Gigatons per year, in 2030 it would be in the atmosphere 520 Gigatons, considering that half of the emission will be absorbed by Mother Nature. And I did not take in calculus permafrost and peatland emission. As we know the weather events of the crazy year 2020, with 420 Gigatons of pollutants in the atmosphere! I ask myself: what would happen with an additional 10 to 20 Gigatons/year in the atmosphere? What would the weather be like? *"That is suicidal for humanity."* The new study shows that the tipping point of 1.5C will be reached before 2025. The impact will be: *"Apocalyptic fires and floods, cyclones and hurricanes are the new normal. Biodiversity is collapsing. Deserts are spreading. Oceans are choking with plastic waste."* What else do we need? Is this not enough? The truth is here, *"without **SACRIFICES**, humanity is doomed"*. As presented in previous chapters, fossil fuel companies are expanding their businesses. The tundra is burning, more GHG in the atmosphere than scientists can digest. Dollar man is strong, influential, and prevais, keeping the business as usual as is. I am sorry to conclude that the *"young generation from today will live and*

die from the disasters of Mother Nature. The generations that are not yet born are close to zero to be born" (216).

In his first week on the job, as the country's first Presidential Climate Envoy, John Kerry has moved aggressively to put the U.S. back at the center of global climate discussions, declaring Washington is *"proud to be back"* at a summit of world leaders. The USA vow to make *"significant investments in climate action"* both at home and abroad. Speaking to Yahoo Finance, John Morton, former White House senior director for energy and climate change at the National Security Council, said: a renewed focus to finance global clean energy projects puts the U.S. in direct competition with Chinese energy investments around the world. *"I think the transition to a global low carbon economy represents the single most predictable and consequential economic transformation in human history,"* said Morton. *"The question is, the speed at which it will occur and who will be the winners and losers from a technology standpoint, a business standpoint, sector standpoint, and indeed... national standpoint. So we should see this as a competitive race, and we should adjust our policies accordingly to see this as a race for the future technologies of the world."* (White House).

The **United States and China.**

Many emerging nations have already set ambitious targets, but lack the necessary funding to deliver them. The Green Climate Fund was set up to help vulnerable countries transition to clean energy. Under former President Obama, the U.S. pledged $3 billion to the U.N. It has only delivered a third of that to date. Meanwhile, *"China has become the largest financier of global energy projects, largely concentrated on fossil fuels"*. An analysis by the Brookings_Institution found that China's largest policy banks, the Chinese Development Bank and Chinese Export-Import Bank, *"financed nearly $200 billion in overseas energy sectors between 2007 and 2016, making China the leading outbound source of funding for electricity generation"*. Fossil-fuel power plants developed with Chinese investments account for roughly 3.5% of global annual carbon dioxide emission (same as total global aviation industry), according to Boston University's Global Development Policy Center. *"China has said they are going to reach carbon neutral by 2060, and pick on fossil fuels consumption by 2025 but we don't* *have a clue yet how they're going to get there. I hope we can work with China, I hope we'll be able to get China to share a sense of how we can get there sooner than 2060,"* John Kerry said, at the World Economic Forum in Davos. *"Those issues will never be traded for anything to do with climate. That's just not going to happen, but the climate is a critical standalone issue,"* Kerry said. *"China is 30% of the emissions in the world, we are 15% of emissions in the world. You add the EU to that, then you have three entities that are more than 55% or so. So we must find a way to compartmentalize and move forward"* (208). As we can see, for these two countries, the climate issue is not only an internal affair, they are interconnected internationally, but have different approaches to the climate crisis. The bottom line is that the Climate Crisis doesn't have borders, no color of the flag, latitude or meridian, is planetary (217).

Europe and The United States

Europe and the United States should join forces in the fight against climate change and agree on a new framework for the digital market, limiting the power of big tech companies, European Union chief executive Ursula von der Leyen said. *"I am sure: A shared transatlantic commitment to a net-zero emissions pathway by 2050 would make net-zero a new global benchmark,"* the president of the European Commission said in a speech at the virtual Munich Security Conference on April 19, 2021. *"Together, we could create a digital economy rulebook that is valid worldwide: a set of rules based on our values, human rights and pluralism, inclusion and the protection of privacy."* (218). Both the USA and The EU have pledged to become a net-zero economy by 2050. A transatlantic alliance could have a positive influence on other big emitters like China, India, Brazilia, Saudi Arabia, and more...to commit to net-zero emission (219).

Putin Orders Russian Government to Try to Meet Paris Climate Requirements.

President Vladimir Putin has signed a decree ordering the Russian government to try to meet the 2015 Paris Agreement to fight climate change, but stressed that *"any action must be balanced with the need to ensure strong economic development."* (This means: First Money, Second Climate Crisis, and to try means on paper and speeches). Russia, the world's fourth-largest emitter of GHG, has previously signaled the wish to reduce pollution up to 70% by 2030 against 1990 level. Is strongly backing his aims on huge tundra and ecosystem, absorbing most of GHG from the atmosphere. For Russia, global warming consists of a huge challenge for its economy based on oil, gas and coal, use and export, as well as mining. Ather big emitter and source of greenhouse gases are India, Middle East countries, Australia, Brazil, Iran, and some countries from Africa and South America, which recently discovered huge petroleum and natural gas reserves. These countries are undeveloped and will extract their resources for development. Countries from Middle East economies are based on the export of fossil fuels, as well as Russia. For all these countries, reducing the extraction of fossil fuels for reaching carbon neutrality by 2050, will be a big issue, and the commitment will remain in words and papers. *"An arms race in the world cost is approximated at $5 billion per day, equivalent to $1.825 trillion per year. This will be enough money to build 2 (CCS), which may absorb 9 Gigatons of CO2 per year,"* that represents 22.5% of global yearly emission. The future generations (hopefully would be such generations), might say about our generation, that they spend trillions on destroying Mother Nature, rather than saving it (Yahoo News UK, 2020).

Direct Air Capture $1 Trillion A Year.

Keeping the planet from warming beyond recognition requires an unprecedented economic effort to halt climate-changing emissions. But even the most optimistic scientific projections call for *"not only stopping the output of carbon dioxide but also removing the vast concentrations of the gas"* already in the atmosphere, where it will continue to trap heat for more than a century to come. New research outlines the importance of the USA's mighty industry to suck the GHG from the air. Known as *"direct air capture"*, (DAC) or *"carbon capture system "*(CCS), such technology could eventually *"extract roughly half the atmospheric CO_2."* Like waging war, doing so will be *"extremely expensive and energy-intensive"*. But the University of California, San Diego, researchers *"set out in their peer-reviewed study."*

The findings offer something of a thought exercise, and a warning: *"Either jump-start the carbon removal industry with generous government support now, or risk playing desperate triage in the hotter, chaotic decades to come".* Two companies are having the technology, *"Carbon Engineering, based in Canada, and Climeworks, headquartered in Switzerland."*

President Joe Biden promised to inject $2 trillion into programs to convert the nation's electricity supply to *"100% zero-carbon sources by 2035". "A fraction of that research funding would go to negative-emissions technologies like (DAC)."* A new study envisions spending about *"$1 trillion per year"* on direct air capture alone. This is roughly 5% of the United States' annual GDP.

By 2050, the study found, a DAC program expanded with that kind of money could remove 2 to 2.5 Gigatons of CO_2 per year. Estimates from the U.N. IPCC vary on how much CO_2 needs to be removed, but most suggest a figure close to 6 Gigatons per year by 2050. (As of 2019 the total emission was at *"40 Gigatons of GHG" = 40 billion tons*). The rest would need to be removed by natural sources. The study points to *"hydropower, nuclear and natural gas as the best options of energies to run DAC plants at present".* The task does not need to fall on the U.S. alone. The European Union, she'll be a partner too *"The U.S. must marshal its allies in the Organisation for Economic Co-operation and Development (OECD), the 37-nation club of rich democracies, to participate in the effort, and **big fossil fuel companies** must have their share in this program. Funds need to be invested today to accelerate the deployment of this technology.* Governments should focus first on zero-carbon sources of energy, adding less GHG into the atmosphere, which could make DAC cheaper. *"Large-scale decarbonization of the power sector, as well as direct and indirect electrification of other industries and technologies, should be prioritized as a cheaper climate change mitigation solution* (OECD, Univ. of San Diego & UN).

A New Study Makes Me Not to Sleep

Even if we stopped emitting greenhouse gases today, the Earth would continue warming for centuries. *"Arctic ice and permafrost are already on an irreversible path of melting and have already reached the tipping points".* The study was published in December 2020, in the journal Scientific

Reports. The model suggests that *"even if emissions were to drop to zero this year"*, by 2021, global temperatures would ultimately *"rise to be 5.4F(3.C) higher in 2500"* than they were in 1850. *"The tundra will continue to melt over the next 500 years, irrespective of how quickly humanity cuts its greenhouse-gas emissions,"* JÃ¸rgen Randers, the lead author of the new study, told Business Insider. That's because *"climate change is a vicious, self-sustaining cycle"*: Permafrost thaws, it releases more GHG, which sustains warming over time. To stop that cycle, Randers said, we'll need to suck carbon dioxide back out of the atmosphere. By modeling various emissions-reduction scenarios on Earth's climate between 1850 and 2500, Rander's study showed that *"if emissions stopped for good in 2020, sea levels in 2500 would still be more than 10 feet (3.05 meters) higher than in 1850"*. To prevent temperature increase, *"greenhouse gas emissions would need to have ceased entirely between 1960 and 1970,"* the model found. *"In that sense, Earth blew by a climactic point of no return 50 years ago"* much before the public understood what climate change is. "Yes, that is an irony," Randers said. *"The scientific community knew about global warming already in the 1960s."* Polar melting is on track to raise seas *"3 feet (0.91 meters) by 2100,"* Randers said that companies and governments need to *"start developing the technologies for large-scale removal of greenhouse gases from the atmosphere."* To prevent further warming after emissions have stopped, the new study found, *"at least. 33 Gigatonnes (36.5 billion tons) of carbon dioxide would need to be sucked out of the atmosphere each year"*. That's roughly the total amount of carbon dioxide the global fossil-fuel industry emitted in 2018 (36 Gigatonnes. At \$100/ton, it means \$3.6 trillion per year).

Note This value of total emission per year, I found to vary between 36-42 Gigatons/year). Power plants in the US, Canada, and Switzerland have already started utilizing CCS to lower their emissions. Two US carbon-capture completed in 2017, *"can capture 2.7 million of tons of carbon dioxide per year"*. The total amount of CO_2 that requires far more plants than any current plans call for. *"In other words, building 33,000 big CCS plants and keeping them running forever,"* the study authors wrote (210).

The Pros and Cons of Geoengineering.

Carbon capture is becoming widely accepted as a safe and potentially effective form of geoengineering. *"Andrew Yang, a 2020 Democratic presidential candidate, suggested budgeting US$800 million for further geoengineering research in the US."* But most *"climate-hacking proposals would be far riskier than CCS"*. Take solar geoengineering, which involves injecting aerosols into the sky to reflect sunlight into space. Critics of this idea point out that most models predict the effects of solar geoengineering wouldn't have borders. *"Aerosol injections deployed in the southern hemisphere, for instance, could impact ocean temperatures and wind speeds, in the northern hemisphere. Solar geoengineering has geopolitical ramifications, unlike carbon capture,"* Juan Moreno-Cruz, an associate professor at the University of Waterloo who studies geoengineering, previously told Business Insider. Randers said his study advocates just for carbon capture, not other more experimental forms of geoengineering. As an immediate priority, he added, countries should invest equally in efforts to cut emissions and build more CCS plants (220). This is the most alarming study I have ever read, but seems to be close to reality.*"These CCS systems need to be financed by Big Fossil Fuel companies* now" not planting trees which will mature in 20-30 years to absorb CO2, too late for any good for the planet.

6.2.4. BIG FOSSIL FUEL CO., INSTITUTIONS, AND INVESTMENTS

New Nonproliferation Fossil Fuels Treaty

The Los Angeles City Council is poised to endorse a call for a *"global Fossil Fuel Non-Proliferation Treaty."* Approval could make Los Angeles the first U.S. city, New York is also in the running, to sign on to the treaty resolution. Introduced in November 2020 by Councilman Paul Koretz, it won unanimous support in committee and awaits passage by the full council in the new year 2021. The treaty would do just what its name says: Signatory governments would agree to stop further expansion of the fossil fuel industry within their boundaries. The treaty *"addresses a nearly universal failing of climate change regulations"*, which usually attempts to curb energy demand, instead of attacking the oil industry directly. The state's most recent two governors have acknowledged the gravity of the climate crisis, *"but their administrations have both issued new oil well permits at a rate of 1,000 to 3,000 a year"*, according to FracTracker Alliance. New drilling permits, and in the first nine months of 2020 jumped to 1,646, an increase of 137% over the same period in 2019. *"The governors' disinterest in stopping new permits is attributable to the continuing power of the oil industry's lobbying and campaign contributions, which exert a strong influence on both parties"*. The *"cynicism of big oil and gas companies"*, not only in the USA but around the planet is outstandingly presented in Mr. Professor Michael Mann's book *"The New Climate War, The Fight to Take Back our Planet"*, published in January 2021. Correspondingly, the industry's claims to its astonishingly large global subsidies, roughly $5 trillion, about 6% of global GDP. Considering fossil fuels' role in destabilizing the global climate, those subsidies are outrageous. Two British social scientists issued the initial call for a global fossil fuel treaty. In a 2018 Guardian op-ed, they proposed the 1970 U.N. Treaty on the Non-Proliferation of Nuclear Weapons, as a model for a similar swift approach to limiting fossil fuel production. Besides committing governments to *"ending the development of new fossil fuel resources,"* the treaty proposal calls for a coordinated, accelerated phaseout of existing fossil fuel production. It would *"divert money from fossil fuel subsidies to clean energy development in poor countries"*. Numerous nations, including New Zealand, France, Costa Rica, Belize, have *"announced moratoriums on new oil exploration and production"*. Others have phased out fracking, and coal production and use. Vancouver, Canada, became the first city to sign on to the treaty. The city of Los Angeles is home to more than *"800 active wells in 26 oil and gas fields"*. In December 2020 a City Council issued an ordinance that could phase out oil and gas production in L.A., *"but any such measure faces heavy industry resistance"*(221).

Big Oil And Gas Company

To make informed decisions about how to allocate capital and manage risks, investors and the public need complete, comparable, and reliable information about companies' climate risks. *"Climate change will impact entire activities for investors, companies, and entire sectors of the economy."* Companies can not regulate themself when it is about climate risk. *"The government must enhance financial regulations"* to inform the markets about the risk posed by climate change. The Securities and Exchange Commission (SEC) has to ensure that financial regulations account for the growing threat of climate change. Some big fossil fuel companies start to voluntarily make disclosures regarding their prime factors and supply-chain risks due to global warming. Public-company disclosures related to climate risks must be standardized and mandated. Now, the new American government should control SEC adoption of several basic reforms. *"While disclosure regulations have eroded across industries, the fossil-fuel industry, in particular, has used the weak regulatory environment to keep investors in the dark when it comes to climate risk".* Earlier this year 2021, Chevron shareholders voted to compel the company to disclose its climate-related lobbying efforts. *"If the SEC required companies selling securities to make basic disclosures related to their climate risks and impacts.* This escape from the public eyes of big fossil fuel corporations is typical and global (222). *"Humanity needs commitments from Big Corporations, to make a dent in global warming."*

Investments And Death

"The world is heading for mortality rates equivalent to the Covid crisis every year by mid-century unless action is taken". But with governments supplying billions into supporting economies, what governments need to assure us is that these investments are green enough.

The pandemic and global warming are lapping, and climate is no *"one-off"* basis. Governments pouring trillions of dollars into stimulus plans are mainly addressing job losses and the economic damage, a very small percentage is directed towards reducing carbon emissions. U.N. envoy Mr. Carney, who is tasked with persuading policymakers, chief executives, bankers, and investors to focus on the environment, said: *"The scale of investment in energy, sustainable energy, and sustainable infrastructure needs to double. This investment will be every year, for the course of the next three decades, $3.5 trillion (£2.5tn) a year.* He said the answer lies in a global amount of *"$170 trillion of private capital which is looking for disclosure".*

That amount of money *"is bigger than the total economic output of the world for a year"* (approx. $120 trillion), he said. Such an investment should be successful as money works better, invest in the right place. As the US returns to climate change, it must work together with China even if China *"draws about 70% of its power from fossil fuel"*, the country is *"a crucial part of the solution"*. In some sectors, China is better positioned for climate change, but the US has the *"largest and most sophisticated financial sector"* along with the "engineering, innovation, and technological expertise" to get to net-zero emissions. Part of the US role at the UN is to tap into this financial sector. *It is the power of money that will ultimately play the biggest role in combating climate change"* (223).

Banks And Investment Co. - Cynicism

"Polluting companies must disclose the full scope of their greenhouse gas emissions or risk confrontation with investors at their annual shareholders' meetings", the world's biggest asset manager (approx. $8 trillion) BlackRock said on Wednesday, February 17, 2021.

For the first time, BlackRock has publicly sought the release of so-called Scope 3 emissions data, so companies account more for climate change. Scope 3 refers to emissions generated by a company's products or during services. For an oil company, this would include the emissions created by running a car, which uses gas from the oil company. That more than *"doubled the number of heavy-emitting companies facing intensive talks on climate risks to 1,000.*

"Where corporate disclosures are insufficient to make a thorough assessment, or a company has not provided a credible plan ... we may vote against the directors we consider responsible for climate risk oversight," BlackRock said (224).

"Institutional investors have $1trillion (£710bn) invested in the thermal coal industry in 2020." New research from 25 climate groups, including Urgewald, Reclaim Finance, Rainforest Action Network, and 350.org Japan, found that *"almost 4,500 institutional investors around the world had $1.03tn invested in the thermal coal"* value chain at the end of the last year 2020 (In these financial institutions are included BlackRock and Vanguard). The estimate was based on an analysis of the Global

Coal Exit List and shows for the first time researchers have sought to measure the entire finance industry's exposure to coal. Almost 60% of the funding comes from US-based investors. *"17% comes from BlackRock, and Vanguard,"* two of the biggest asset management companies in the world. Researchers said the two investment companies were *"in a class of their own."* What you say and what you do never connects.

"Also, the banking industry's support for the coal sector has been analyzed." 381 banks have lent $315bn to the coal industry in the last two years 2019 and 2020. *"Commercial banks"* have also helped the sector raise over *"$800bn through bond issues and share sales".*

The three biggest *"coal lenders were based in Japan, but Citigroup and Barclays ranked fourth and fifth respectively".* Both have lent over $13bn to companies involved in coal. What banks and asset managers did, were not in line with the goals of the 2015 Paris Agreement.

"The bulk of coal financing and investment must end NOW." Many professional investment companies and banks are taking steps to address climate issues. BlackRock founder and chief executive Larry Fink last year 2020, pledged to make tackling the climate crisis his company's top priority. He said sustainability would be BlackRock's *"new standard for investing"* and repeated his pledge last month in February 2021. *"His words speak his mind and conscience".* Banks like Citi and Barclays have also made commitments to transition to net zero. Citi has publicly committed to reducing its exposure to coal companies to zero by 2030. *"These numbers provide a sobering reality check on the bank's climate commitments,"* said Yann Louvel, a policy analyst for Reclaim Finance. *"The vast majority of banks' coal policies have so many loopholes that their impact is almost meaningless."* (Yahoo Financial News, 2020).

G20 Subsidies For Pollution

Rich nations are still providing "more than half a trillion dollars annually to fossil fuel projects" despite committing to slash GHG emissions per the Paris agreement, research showed in November 2020. A joint analysis by three climate think tanks found that many nations' *"post-pandemic stimulus plans will see billions more given to polluting fuels".*

In grading G20 countries' performance of phasing out fossil fuel subsidies, the analysis found that *"at least $170 billion of public money had been pledged to fossil fuel-intensive sectors"* since the start of the pandemic (less than a year). *"G20 governments were already not on track to meet their Paris Agreement commitments on ending support for fossil fuels before Covid-19,"* noted Anna Geddes of the International Institute for Sustainable Development, which co-authored the report. *"Now, disappointingly they are moving in the opposite direction."* The International Energy Agency said that the pandemic provided a once-in-a-generation opportunity for governments to enact a *"step-change in clean energy investment"*. But administrations desperate to get their economies firing again after the damage wrought by Covid-19 seem not to have taken this advice to heart. It found they had given $584 billion to fossil fuel projects every year between 2016-2019. Gauging each country's subsidy, the analysis found that not a single nation has made *"good progress"* in line with the Paris accord.

A new trend for big oil and gas companies is to make pretty large investments in renewable energy. That is a good investment but does not solve the emission of greenhouse gases. We are at the beginning of the second decade of the 21st century, and the least is done by governments, the economy, the public, and others contributing factors to the climate crisis.

An important factor is "education", on all levels and branches of society (225).

Earth Day, April 22, 2021

On Earth Day in this year 2021, the President of the United States has sponsored a virtual summit to at least 40 chiefs of states or governments, to present the urgency of the climate crisis. President Biden's plan calls for eliminating GHG emissions from the power sector by 2035, and from the entire economy by 2050. Biden has committed to *"at least a 50% reduction in U.S. emissions by 2030"*, compared with 2005 levels. Biden's plan will involve *"massive increases in renewable forms of energy such as solar and wind power"*. But just as important will be *"revamping the U.S. electrical grid"* so that clean power can get to people far away from energy sources. Mr. John Kerry, mentioned that direct air capture from the atmosphere is a requirement if society must reduce global warming, saying *"that almost 400 billion tonnes of GHG shall be captured, and sequestrated. (This is*

almost the entire quantity of GHG existing nowadays in the atmosphere)."
He did not mention any period of time, or how much per year? Anyway,
it is important that he raised this detrimental issue, as presented in ante-
paragraphs, and is a most discussed aspect of the climate crisis in the
scientific circle. China has agreed that the country will reach net zero by
2060. No specification for short-term achievements. This reduction can be
done linearly or parabolic. If parabolic, which means all percentages to be
reached in the last 5 or 10 years (as discussed), then the reduction for the
climate crisis, *"equals to ZERO effect".*

President Vladimir Putin said he wanted Russia's total net greenhouse
gas emissions to be less than the European Union's over the next 30
years, a goal he described as tough but achievable. Russia is the world's
fourth-largest greenhouse gas emitter. *"This is a difficult task, given the
size of our country, its geography, climate, and economic structure. However,
I am absolutely certain that this goal, given our scientific and technological
potential, is achievable."*

*Putin has said Russia is warming at 2.5 times the world average and
that it would be a disaster if the permafrost melts in its northern side of
Russia.* If Russia's reduction of GHG will take the same shape as China's
parabolic curve, we can conclude that the effect on climate emergency will
be undetectable. Mr. Putin wants credit for his country's vast Taiga forest,
which he claims absorbs more carbon that Russia is releasing. Perhaps
Mr. Putin forgot that his tundra permafrost last year 2020 was burning
for more than 3 months and that the permafrost and northern peatlands
contain almost the same amount of GHG as it is in the atmosphere
today (400 Gigatonnes). Brazil's president may cut deforestation in the
Amazon rainforest up to 40%, if the USA is helping with a donation of
$1 billion. Brazil's President Jair Bolsonaro approved a 24% cut to the
environment budget for 2021 from last year's level, according to official
numbers published on Friday, April 23, just one day after he vowed to
increase spending to fight deforestation. Speaking on Thursday, Earth
day, April 22, to the summit Bolsonaro pledged to double the budget for
environmental enforcement and end illegal deforestation by 2030. With
this kind of leader, humanity is fighting the climate crisis. India and other
big emitters did not present any noticeable proposal of cutting the emission
of GHG. I can conclude that this summit was a prelude to the coming up

of the International Panel of Climate Change (IPCC) conference, from 1 to 12 November 2021, in Glasgow, United Kingdom (225A).

6.2.5. EDUCATION VS. DENIAL

"Extraordinary People Fighting to Save Our Ailing Planet"
"Right now, the world is at a defining moment in the fight against a climate catastrophe".

And while it's easy to perceive our current situation as bleak, hope is still out there to slow down gigantic waves. Everyone called humans has to participate as shown by the *"extraordinary young people who are doing the best to save the planet.* Snap Inc. uses Spectacles to tell stories through the eyes of its subjects. Hosted by *"mobile journalist, Yusuf Omar,"* the series documents the daily lives of inventors, professionals, activists, and citizens as they actively present climate consequences of warming and environmental destruction through their work. The stories and work of these people may have an impact on people on the other side of the plane. Their action tells the present generation that the young generation cares about this planet on which they and their kids will live. The question is: *"Is it not too late? To wait for their (young generation) action and we (mature generation) to do nothing to save the planet"?* (226).

Can We Educate Giant Corporations?

Before Jonathan Scott's new documentary about clean energy premiered, the home renovation star revealed what inspired him to advocate for making solar energy accessible for all. Scott first was working on his house in Las Vegas installing solar panels. Performing on his project he finds out *"the local utility had all of these crazy rules that limited how much solar [he] could install and how long it would take to activate."* He succeeded in getting the system activated, then the *"local utility convinced the public commission to stop net metering in Nevada."* This *"essentially put all the solar companies out of business."* Scott told himself: now is the moment to know more. He found that other states had it even worse. *"There was a concerted effort to stifle solar innovation and prevent people from harnessing the power of the sun for themselves."* He says that *"when he found out that fossil fuel* utilities were spreading misinformation and learned more about the negative effects of fossil fuel on low-income communities that lack resources to fight back, he decided to take action." Scott met with people involved in the fight for clean energy. *"When I spoke with people in neighborhoods that had been polluted and endangered by utility companies and then were told they would charge those same residents for the cleanup, that made my blood boil,"* he told People. *"Everyone deserves the opportunity to live in a safe, healthy community without the fear that some giant company will poison them and make them pay to fix it."* (227). This kind of journalism teaches the *"big fossil fuel companies to see in their mirror the cynicism and stupidity they are engaged in."*

Practical And Surprisingly Uplifting Guide to Living With Climate Change.

In *"How to Prepare for Climate Change"*, David Pogue, a five-time Emmy Award-winning technology and science correspondent for CBS Sunday Morning, New York Times bestselling author, and host of science specials on PBS NOVA–offers a sensible, thoroughly researched guide with practical advice for how everyday people can react themselves for the tumultuous years ahead. Climate change started long ago, around 1960 if you might not realize it. Climate Crisis if it starts can not slow down for centuries. And we already are in this crisis. It's too late to stop it. Now it's time to adapt. With the help of 50 experts in different fields of study, author David Pogue brings his trademark clarity, authority, and even humor to an unexplored subject: *"preparing your family, your home, your business, and your life for our planet's chaotic crisis future"* (228). We Are Welcome to Planet Earth.

Cynicism And Denial. What we need to fight via EDUCATION
"The five corrupt pillars of climate change denial"
Mark Maslin is a Founding Director of Rezatec Ltd, Director of The London NERC Doctoral Training Partnership. He analyzed the role of fossil fuel companies versus climate change.

"The fossil fuel industry, political lobbyists, media moguls, and individuals" has spent no time and money in the last 30 years, presents false scientific information to inflict doubt about the reality of climate change on the public, so climate change does not exist. The world's biggest public-owned

fossil fuel (gas and oil) companies are spending annually US$ 200 million on lobbying *to control, delay or block binding climate policy."*

They can not go in this direction. Two recent polls suggested over *"75% of Americans think humans are causing climate change".* School climate strikes, Extinction Rebellion protests, UN and national governments declaring a climate emergency, dynamic improved media coverage of climate change, and extreme weather events have all contributed to this shift and changed the public perception about global warming. The big companies have changed the lobbying, now employing more subtle and more vicious approaches, in their view climate change is climate sadism. Nowadays it is important to identify the different types of denial for climate change. The below types of denials presentation will help *"you, the reader",* spot the different ways that are being used to convince you to delay action on climate change.

1. Science - denial.

This is the typical type of denial: that the science of climate change is not known enough and settled. Deniers suggest climate change can not be controlled and is part of the natural cycle. Also, climate models are not well defined and too sensitive to carbon dioxide. CO_2 is such a small part of the atmosphere, how can it have a large heating effect? One more argument is that scientists are fixing the data to prove the warming (a global conspiracy that would take thousands of scientists in more than 100 countries to pull off). Denial's arguments are smelling from far away, they are false, and scientists around the globe have a clear consensus about the causes of climate change. All science data are in accordance with weather phenomena and shifting in public opinion pro-climate change. A new tactic for denial for instance is Britain's leading denier, Nigel Lawson, who is able to say that: humans are causing climate change. It says it is *"open-minded on the contested science of global warming, [but] is deeply concerned about the costs and other implications of many of the policies currently being advocated".* In other words, climate change *"is now about the cost, not the science."*

2. Economic denial

"Economists suggest we could fix climate change now by spending 1% of world GDP, even less if we take into consideration savings from improved human health and benefits of the global green economy are taken into account. If we don't do anything, by 2050 it could cost over 20% of world GDP." (Note: The author of the article is using the word *"fix"* for the climate. As shown in chapter 2, there does not exist the notion *"fix"* for the climate crisis, there may be *"a slowdown or attenuate the impact of global warming"*). We should also remember that in 2018 the world generated US$86. trillions and annual rates of growth of global GDP is 3.5%. One percent of global GDP will be an easy way to treat the issue of the climate and would save the world a huge amount of money. The deniers also forget to tell us that *"they are protecting a fossil fuel industry that receives US$5.2 trillion in annual subsidies"*. This amounts to *"6% of world GDP."* The International Monetary Fund (IMF) estimates that *"efficient fossil fuel pricing would lower global carbon emissions by 28%, fossil fuel air pollution deaths by 46%, and increase government revenue by 3.8% of the country's GDP."*

3. Humanitarian denial

The deniers also, full of cynicism, argue that climate change is good for us. They say that longer, warmer summers in the temperate zone will increase farming productivity. Warmer summer means drier summer and increased frequency of heatwaves. In the *"2010 Moscow heatwave killed 11,000 people, devastated the wheat harvest, and increased global food prices."* For the Tropics where lives approx.n "40% of the world's population, both from a human health perspective and an increase in desertification are increasing the health cost and productivity. Deniers also say that more atmospheric carbon dioxide is better for plants and forests. This is indeed true up to one point, and the *"land biosphere has been absorbing about 25% of our carbon dioxide pollution every year"*. Another quarter of our emissions is absorbed by the oceans. Also, climate deniers will tell us that more people die of the cold than heat, so we benefit from warmer winters. Poor communities experience poor housing, during wintertime they experience cold difficulty. This argument is also factually incorrect. In the US, for example, *"heat-related deaths are four times higher than cold-related ones."* Many heat-related deaths are recorded by cause of death such as heart failure, stroke, or respiratory failure.

4. Political denial.

If other countries are not taking action, why should we take action? The deniers suggest. Not all countries contribute equally to CO2 emission. For example, 15% of CO_2 in the atmosphere is generated by the US, another 10% is produced by the EU, China 28%. Africa produces just under 5%. Given the historic legacy of GHG pollution, EDC countries have an ethical responsibility to be leaders in cutting emissions. Finally, all countries have to participate together in combating the emission of GHG. The effect of global warming does not have borders, so should be the actions. Deniers will also tell us that we need to solve our own issues for our country, not waste time with global issues. Solutions to climate change are win-win for all participants and will improve the environment, food, and water security, and reduce pollution.

5. Crisis denial

Deniers argue that we don't need to change things. Deniers are emphasizing that climate change is not as bad as scientists present it. Using fossil fuel we can be richer in the future and better able to fix climate change. Based on the last 2 years of climate events, the deniers can not argue anymore that we are better and in the power of our destiny. In the past have been used arguments to delay ending slavery, granting the vote to women, ending colonial rule, ending segregation, decriminalizing homosexuality, bolstering worker's rights and environmental regulations, allowing same-sex marriages, and banning smoking. The fundamental question is *"why are we allowing the people with the most privilege and power to convince us to delay saving our planet from climate change?"* (228).

Global 'Elite (Rich)' And High-Carbon Lifestyles.

"The world's wealthiest 1% account for more than twice the combined carbon emissions of the poorest 50%", according to the UN. The richest must rapidly cut their CO2 footprints to avoid friction with the rest of the population, and dangerous warming this century.

The global Covid-19 shutdown will have an insignificant impact on the climate. The study has been compiled by the UN Environment Programme (UNEP). The report *"examines the roles of lifestyles and consumption patterns of individuals."* The global top 10% of incomers use around 45% of all the land energy and around 75% of all the energy for aviation, compared with just 10% and 5% respectively for the poorest 50%. These high *"carbon footprints" must be "curbed to around 2.5 tonnes of CO2 per capita"* by 2030. *"But for the richest 1%, it would mean a dramatic reduction".* The *wealthy bear the greatest responsibility in this area."* The combined emissions of the richest 1% of the global population account for more than twice the combined emissions of the poorest 50%. *"This elite must reduce their footprint by a factor of 30 to stay in line with the Climate Change targets."* These types of lifestyle changes will improve everyone's life. The UNEP report says that not too much has been done to curb the pollution, and more action needs to be taken to meet net-zero emissions by the middle of the century (229).

I need to emphasize that mass media, schools, and universities have a tremendous impact on the climate education of the masses. As far as I know, there do not exist any courses or classes (seminaries) taught in the educational system on the theme of climate change. Also, wish to praise the American channel news CNN for outstanding programs with Climate Change on the subject. I hope it will continue and other channels will follow. Also, the faith leader can bring to the masses their message on climate crises. Here to mention repeated messages by Pope Francis and Dalai Lama.

7

CONCLUSIONS

"I've always been terrified for the future of young people, in particular, but it feels so much more real now. This is the kind of stuff I thought they would be dealing with, and now we're dealing with it. So what does that mean they are going to be dealing with in 10 years or 30 years? I can't even imagine." (Peter Kalmus). And I am terrified for those *"generations who are not yet born"* (6). Hundreds of scientists, writers, and academics from 30 countries sounded a warning to humanity in an *"open letter"* published in the Los Angeles Times on December 31st, 2020. Policymakers and the rest of us must *"engage openly with the risk of disruption and even collapse of our societies"*. Overloaded atmosphere with GHG and deterioration of plants and animal's environment will be the cause of *"social collapse" as "a credible scenario this century."* Future Earth group has surveyed a pool of scientists and found that extreme weather events, and their consequences on food insecurity, freshwater shortages, and the broad degradation of life-sustaining ecosystems *"have the potential to impact and amplify each other and create a domino cascade effect. That effect will automatically imply a global systemic collapse."*

The Breakthrough National Center for Climate Restoration, a think tank in Australia, found that a warming climate and increased pollution will cause a global depletion of resources. Such conditions would lead to *"a largely uninhabitable Earth,"* and a *"breakdown of nations and the international order."* At the same time, analysts from the UK and the USA military analysts have issued for two years threats of global warming.

For nonhuman species, *"collapse is well underway: 99% of the tallgrass prairie in North America is gone, 96% of the biomass of mammals, now consists of humans, our pets, and our farm animals"*; nearly *"90% of the fish stocks the U.N. monitors are either fully exploited, overexploited or depleted."* Note: biomass is their weight on Earth. A study in Germany determined that there is a 76% decline in insect biomass. *"This is a very drastic and alarming situation"*. The pandemic brought the public engagement in climate issues to NONE, determining in some technologically advanced nations: uncontrolled pandemic, institutional failures, and economic and labor issues. Among the scientists signing the warning letter were:

a). William Rees, a population ecologist at the University of British Columbia best known as the originator of the *"ecological footprint"* concept (different from carbon footprint), which measures the *"total amount of environmental input needed to maintain a given lifestyle"*.

The society's footprint of energy and some key resources for technological development leading to *some form of global societal collapse is inevitable, in the next decade, certainly within this century,"* Rees said in an email. *"The biophysical collapse is what he calls overshoot: humans exploiting natural systems faster than the systems can regenerate"*. Humanity's technical development has driven society to prosper, but causing *"liquidation of biophysical capital essential to its own existence.* What humanity dumps Mother Nature does not have the capacity to assimilate. All impacts of climate change, like pollution, ocean acidity, deforestation, among other problems, are very important on their own. Each of them is a mere symptom of overshoot, says Rees. The message we should take from tons of evidence is that human development is ultimately determined by biophysical limits. *"We are exceptional animals, but we are not exempt from the laws of nature"*.

b). Will Steffen, a retired Earth systems scientist from Australian National University. Steffen put in evidence the *"neoliberal economic growth paradigm"*, the more and more increase of the GDP, as *"incompatible with a well-functioning Earth system at the planetary level."*

Collapse, he said, *"is the most likely outcome of the present trajectory of the current system, as prophetically modeled in Limits to Growth."* Limits to Growth is a *"150-page bombshell of a book"* published in 1972. *"The authors, a team of MIT scientists"*, created a computerized system-dynamics model called **World3,** the first of its kind, to examine worldwide growth trends from 1900 to 1970. They studied historical data to model *"12 future scenarios projected to the year 2100."* The results showed that any system based on *"exponential economic and population growth crashed eventually"*. The funniest model was the one in which the *"present growth trends in world population, industrialization, food production, and resource depletion continue unchanged,* (as is even today). In that *"business as usual scenario, the collapse would begin slowly in the 2020s and accelerate thereafter."* Note: the authors of the study were goddamn so correct, proved by nowadays reality. Let's take only *"discussing the consequences of our biophysical limits, the December warning letter says:* can we reduce their likelihood, *speed, severity, and harm."* And the authors of the letter are likely to be: *"ignored, crowned doomers, collapses, marginal and therefore discountable"*. We wish everything to be prosperous and well. *"Man is a victim of dope/In the incurable form of hope,"* as poet Ogden Nash wrote. The authors are hopeless for improving the actions on climate. The scientist asked humanity to look directly into the abyss of collapse. Can we deal with what we see there?

Note: Christopher Ketcham is the author, most recently, of *"This Land: How Cowboys, Capitalism, and Corruption Are Ruining the American West."* Jeff Gibbs is the writer and director of the documentary *"Planet of the Humans."* Hope this letter has given you *"The reader"* a second sought about *"Life on the planet Earth"* (218).

When you look at images of the bushfires in Australia or the cracking ice shelves in Antarctica, it's easy to think that *"it's too late to do anything about the climate crisis"*, that we are, for all intents and purposes, **in a catastrophic danger zone**, (not to use American expression f----d). **And it's true,** *"it's too late for 182 people who died from exposure to extreme heat in Phoenix in 2018, or for 1,900 people in northern India who have swept away in extreme floods in 2019, or the 7.6 million people (even more) who die each year around the world from particulate air pollution caused by our dependence on fossil fuels"*. And the way things are going, it's probably too late for the glaciers on Mount Kilimanjaro, for large portions of the Great Barrier Reef, and for the city of Miami Beach as we know it. *"But the lesson of this is not that we're doomed, but that we have to fight harder for what is left.*

Too Late-ism *only plays into the hands of Big Fossil Fuels"*, and all the activists who want to drag out the transition to clean energy as long as possible. Too Late-ism also misses the big important truth that, buried deep in the politics and emotion of the climate crisis, you can see the birth of something new emerging. *"The climate crisis isn't an 'event' or an 'issue,'"* says futurist Alex Steffen, author of **Snap Forward**, an upcoming book about climate strategy for the real world. *"It's an era, and it's just beginning."* This era like any other era takes centuries, if not more, to be back to normality. According to a new poll from the Yale Program on Climate Change Communication, *"nearly six in 10 Americans are now alarmed or concerned about global warming"* (17). Germany, the industrial powerhouse of Europe, *"plans to shut down all coal plants by 2038. Too late to help society attenuate the Climate Crisis"*. In the U.S., the coal industry is in free-fall. Larry Fink, the CEO of BlackRock, the financial giant that manages over $7 trillion in assets, acknowledged in a letter to shareholders that climate change is now *"on the edge of a fundamental reshaping of finance."* Jim Cramer, CNBC's notoriously cranky Wall Street guru, said in January 2020, *"I'm done with fossil fuels. ... We're in the death-knell phase. .... The world has turned on them. It's kind of happening very quickly."*

   "I don't have any doubt that we will take action on climate," says Steffen. *"But it won't be the old-fashioned version of social change. It won't be an orderly transition. It won't be the climate version of the civil-rights movement. It will be more like the Industrial Revolution, a huge social and cultural, and economic transition, which will play out over decades, and with no clear leadership and nobody in control."* Imagine a world adapting entire infrastructure to meet the new requirements of climate change. Same for the economy, banks, and financial institutions, industry, education system.... The countries which will delay this transition process shall be left behind. For instance, in the USA there is a big fight between democrats and republicans to get the new deal for updating infrastructure to be compatible with climate change like China does. Republicans are addicted to fossil fuels. I hope it won't be too late to wake them up. On top of that, we have to add the change in social behavior at work or private places. How we eat, dress, illuminate the places we leave, how we build,...

   In Steffen's view, *"climate doomers are as blind as climate deniers. The apocalyptic is in its very heart a refusal to see past the end of an old worldview, into the new possibilities of the actual world"* (18). The UN secretary-general has called on all countries to declare a *"climate emergency."* António Guterres was speaking at a virtual summit on the *"fifth anniversary of the Paris climate agreement"* on December 12, 2020. He criticized rich countries for spending 50% more of their pandemic recovery cash on fossil fuels compared to low-carbon energy.

Over 70 world leaders are due to speak at the meeting organized by the UK, UN, and France. Mr. Guterres said that *"besides EU countries 10 more countries had already"* declared a climate emergency. *"Emergency would only end when carbon neutrality was reached"*. On the Covid recovery spending, he said that this is money being borrowed from future generations. *"We cannot use these resources to lock in policies that burden future generations with a mountain of debt on a broken planet. Many big emitters, including Australia, Saudi Arabia, Russia, and Mexico, are not taking part"*. The UK has announced an end to support for overseas fossil fuel projects. UK Prime Minister Boris Johnson said: *"Together we can use scientific advances to protect our entire planet, our biosphere, against a challenge far worse, far more destructive even than the coronavirus. And by the Promethean power of our invention, we can begin to defend the Earth against the disaster of global warming.* President Macron of France, Pope Francis will also address the meeting. The UK pointed to its new commitment to overseas fossil fuel projects as well as a new *"carbon-cutting target of 68% by 2030"*. The *"EU also presented a new 2030 target of a 55% cut in emissions"*, agreed upon after all-night negotiations. *"That is now Europe's calling card,"* said Ursula von der Leyen, President of the European Commission. *"It is the go-ahead for scaling up climate action across our economy and society."* China's President Xi Jinping reiterated a previous commitment to reach peak CO2 emissions before 2030 and *"achieve carbon neutrality, where any emissions are balanced by removing an equivalent amount from the atmosphere, by 2060"*. He announced that China would reduce its carbon emissions per unit of gross domestic product (GDP) by over 65% compared with 2005 levels. The country will also increase the share of non-fossil fuels in primary energy consumption by about 25%. And President Xi pledged to increase forest cover and boost wind and solar capacity. *"These measures are too far to be enough to reduce the impact of climate change. China has to do more, as the top emitter of GHG in the world."*

Narendra Modi, the Prime Minister of India, said the country's renewable energy capacity was on target to reach 175 Gigawatts by 2022, and it would aim to boost this to 450 Gigawatts by 2030. *"For a country of over 1.4 billion population, these targets are like a drop in a bucket".*

Pope Francis said the Vatican had committed to reaching net-zero emissions, similar to carbon neutrality, before 2050. *"The time has come to change course. Let us not rob future generations of the hope for a better future,"* he said. Australia had held out the promise of not using old carbon credits to meet future cuts in emissions. Sounds like *"sick-minded behavior."* Australian Prime Minister Scott Morrison, like other leaders, will be on the world list of climate criminals against humanity. Russia, South Africa, and Saudi Arabia won't be involved either.

"UK experts play a central role in the Intergovernmental Panel on Climate Change, advising on science and policy. Where science is conscience too". On economics, the 2006 UK Stern Review showed that *"ignoring climate change was more costly than tackling it. From a symbolic procedural point of view, it's good to have everybody on board,"* said Prof Heike Schroeder from the University of East Anglia. But from a proactive, creating some kind of sense of urgency approach it also makes sense to say we *"only get to hear from you if you have something new to say"*(UN Report). The introduction of the book has presented a summary of all .... chapters. I emphasize the role of *"science and technology"* in the development of society, and the stress they impose on the planet's environment. Scientific research in weather changes must be funded and sponsored by multiple countries or multinational consortia. These multinational cooperation are enhancing the *"science and technology developments with a "global aspect".*

Doomsday Clock set to *"two minutes to midnight"* due to climate change and nuclear weapon threats to humanity. How close are we to Armageddon? The Bulletin of the Atomic Scientists moved the hands of the symbolic *"Doomsday Clock to 100 seconds to midnight".* It's the closest since America and the Soviet Union detonated their first hydrogen bombs in the 50s. The Bulletin of Atomic Scientists said the move was due to two specific threats and described the world's current situation as "the new abnormal". These major threats of nuclear weapons and climate change were exacerbated this past year by information warfare. The new *"abnormal"* that we describe, and that the world now inhabits, is unsustainable and extremely dangerous. The Bulletin was founded by concerned US scientists involved in the Manhattan Project that developed the world's first nuclear weapons during the Second World War. In 1947 they established the Doomsday Clock to provide a simple way of demonstrating the danger to the Earth and humanity posed by nuclear war (6). In the year 2019, Bulletin official Lawrence M. Krauss said, *"To call the world nuclear situation dire is to understate the danger and its immediacy. The same is true with the climate crisis today. Even if human-caused greenhouse gas emissions can be reduced to zero, global temperatures will continue to rise for centuries afterward",* according to a scientific study published Thursday, November 12, 2020. "The world is already past a point of no return for global warming," the study authors report in the British journal Scientific Reports. The only way to stop the warming, they say, is that *"enormous amounts of carbon dioxide have to be extracted from the atmosphere."* The scientists modeled the effect of greenhouse gas emission reductions on changes in the Earth's climate from 1850 to 2500 and created projections of *"global temperature and sea-level rise. According to our models, humanity is beyond the point of no return when it comes to halting the melting of permafrost using greenhouse gas cuts as the single tool,"* lead author Jorgen Randers, a professor emeritus of climate strategy at the BI Norwegian Business School, told AFP. *"If we want to stop this melting process we must do something besides, for example, suck carbon dioxide out of the atmosphere and store it underground, making Earth's surface brighter,"* Randers said. *"The study said that by the year 2500, the planet's temperatures will be about 5.4F (3C) warmer than they were in 1850. And sea levels will be roughly 8 feet higher (2.45 meters)".* To prevent temperature and sea-level rises, *"all human-caused greenhouse gas emissions would have had to be reduced to zero between 1960 and 1970. We need to do in this decade 2021 to 2030, what we should have done 50 years ago. To prevent global temperature and sea level rises, at least.*

33 gigatons of carbon dioxide would need to be removed from the atmosphere each year from 2020 onward through carbon capture and storage methods". (19).

As shown below, this sucking of carbon from the atmosphere is technologically impossible nowadays. Expert Mark Maslin, a professor of climatology at University College London, pointed to shortcomings in the model, telling AFP that the study was a *"thought experiment. What the study does draw attention to is that reducing global carbon emissions to neutral zero by 2050 is just the start of our actions to deal with climate change,"* Maslin said.

In these first 4 pages of Conclusion, I present the scientific clear view of the outcome of squeezing the planet of its resources, like humans are the owner of the Planet Earth and not the guest of it. Also, what the UN has to say in the fight with the climate crisis is that *"this is a global phenomenon, not the issue of a group of countries or a region of the planet."* All the countries must participate and take part in this process. Through global warming, factors, and numbers I look with an engineering mind, mine mind. The developed and developing countries hungry for their industry are consuming the energy produced by burning fossil fuels. These are the sources of greenhouse gases (GHG) in the atmosphere, which produce overheating of the planet. We need to face and fight fossil fuel burning and the electricity obtained through burning. Based on UN data, the average emission of GHG per year is 42 Gigatons (42 billion tons). The existing GHG in the atmosphere is 420 Gigatons (420 billion tons), 10 times more than yearly release. How much energy is produced by *"burning fossil fuels?* Fossil fuel energy represents 82.3 %, versus renewable including nuclear and hydro 15.7 %, which is 5 times less than fossil fuel share. The existing GHG, and what was released in 2019, triggered the weather events in 2020, presented in chapter 2. To slow down the weather events, we need to reduce the existing GHG in the atmosphere, not to release more. Science says that half of the release is captured by oceans and ground vegetation. To note the deforestation and acidity of the oceans are diminishing the absorption of GHG. The *"permafrost and peatland are burning and thawing, releasing more carbon dioxide and methane, which is not considered in that 42 Gigatons"*. Only *"permafrost and peatland store 415 Gigatons of GHG, almost equal to the amount of GHG existing in the atmosphere"*. If we consider that half of the gases released are absorbed by Mother Nature then the other half of 21 Gigatons is getting in the atmosphere and increasing the effect of global warming. In other words, the quantity of GHG in the atmosphere is increasing each year with a minimum of 21 Gigatons. Based on the facts, society has to suck *"minimum 21 Gigatons of carbon out of the atmosphere yearly."* As presented above, some scientists propose 33 Gigatonnes of GHG be captured from the atmosphere yearly.

Let's see how **green energy** stays versus oil and gas energy.

The stone age didn't end due to a lack of stones, and the oil age will end long before the *"world runs out of oil"*. This quote, often attributed to Saudi Oil Minister Sheikh Yamani, highlights a vital and frequently misunderstood fact about the oil industry. *"Oil supplies are not going to run out, but oil will eventually be replaced by cheaper, cleaner, and more efficient energy sources."* This misunderstanding has led many analysts to predict the death of the oil industry.

The *"achieving and adoption of green energy (17.3% of total energy) at incredibly unrealistic speed, is impossible"*. Secondly is ensuring that the existing grid system that we are transitioning to does what it needs to do. It is important to note that *"we do not currently have many of the technologies that we will need if we are to reduce carbon emissions to maintain a leveled atmospheric concentration of GHG"*. We must analyze the carbon capture and sequestration (CCS). Most energy transition plans (CCS) systems or direct air capture (DAC) are some of the key technologies that will help curb CO_2 emissions by capturing carbon from the environment and storing it underground. However, as of 2020, *"only 26 facilities were working globally, capturing 40 million tons of carbon dioxide"* (a thousand timeless less than emitted). Meanwhile, in *"2019 the world emitted more than 39 billion tons of carbon"*. Closing that gap requires time, (and we are out of time), and serious technological breakthroughs, which may be why Elon Musk is offering $100 million to the best carbon capture technology in his new competition. The quantity of carbon that has to be captured yearly varies from group to group of scientists and is between 3 to 33 Gigatons. In a perfect world, a minimum of 21 to 30 Gigatons would be outstanding, but it is impossible. The oil and gas industry has a huge infrastructure, including pipelines, wells, and other facilities. This is the infrastructure that our current energy grids are based upon. For instance, the U.S. would have to dedicate 25–50% of its landmass to solar, wind, and biofuels if it hopes to cover U.S. energy consumption with renewables. Oil and gas companies will oppose these drastic changes and will take significant time and investment. The next big problem with the energy transition is the inherent limitations of renewable energy sources. One of these is power density. The power density of fossil fuel energy is *"two to three orders of magnitude"* higher than renewable energies. Renewable energies, due to their low power density, require vast areas of land. The

USA needs *"33,000 square kilometers of land* with solar panels to cover its demands of total energy. Similarly, it would require using *"half of the UK landmass for wind turbines to cover the country's demand with wind power."*

It is true, *"energy transition is here, the gap between theory and reality here is vast."* Fossil fuels supply almost 83% of global primary energy needs in 2019. According to the Climate Action Network, the EU will need to increase its use of renewables by 50% by 2030 and 100% by 2040 to try to attenuate the climate crisis. While optimism has helped to galvanize support for a global energy transition, a degree of realism is necessary as well, but we need to look at the real numbers of the energy sector. As we look above these numbers are more dramatic and scary regarding renewable energy. It will likely be decades before a full global energy transition can take place, and in the meantime, the most effective way to reduce emissions would be *"to control both ends: production and consumption"*. Degrowth would be a reasonable response to the climate challenges we face, but the pain would outweigh the benefits, especially for the world's poorest. That means financial, technological, and social difficult hurdles, *"and predicting how long it will take is a near-impossible task."* I want to underline that *"France and Japan support nuclear energy"* using new updated technologies and small reactors, which received total attention from America to Arab countries. For instance, Duke Energy isn't going to reach its carbon-cutting goals without the help of a controversial source: nuclear power. Nuclear power is relatively expensive and some environmental groups consider it risky. It's the only source of carbon-free high-density power that can run around the clock. *"The safety record in the U.S. is extraordinarily strong."* It is impossible for the USA to reach carbon goals without *"nuclear-being part of the equation."* Another aspect of carbon emission is the permafrost in Russia, Greenland, and Alaska, which has burned for months in 2020. If this process continues in 2021 scientists consider that the tipping point for this phenomenon is reached. The science considers that 3 more tipping points have already been reached, the melting of both polar ices and the warming of the Atlantic ocean surface's water (changed Atlantic currents).

I insisted on these two factors of the climate crisis: *"emission of carbon, and sources of green energies,"* to show that *"We Are Doomed"*. Even more fossil fuel companies, as shown in chapter 3, *"are not reducing"* the extraction and consumption, they are looking for new reservoirs to expand their businesses in all continents. The thirst for higher GDP in both developed and developing countries causes an increased demand for primary fossil energy. Asia is addicted to coal. We need to reduce extraction, but the reality is the other way. On top of this addiction stands China. Based on the data on China's next 5 years of development, the use of coal will grow at least until 2030. In other words, the quantity of GHG emitted in the atmosphere will not decrease but will increase, in reference to the 2019 level.

Big banks and investment institutions are funding trillions of dollars in fossil fuels. *"The world's largest banks have funneled $3.8 trillion into the fossil fuel industry over the last five years"*, according to new figures published on March 24, 2021, fuelling chaos in the climate crisis. Governments like *"Russia, China, Australia, Saudi Arabia, Mexico, Brazilia are all "in red" for climate change actions"*. For instance, in Russia, the percentage of renewable energy is around 1% of total country energy consumption. *"The present situation of facts does not look good, and is very concerning."* The actions presented in the previous paragraphs, green energy expansion, energy efficiency, energy consumption in building, transportation, aviation shipping, metallurgical industries, electric cars, reducing carbon footprints for products, incentives cuts, border, and carbon tax, elit people footprints cut, and so on, will help, but will not slow down global warming. The attitude of oil and gas companies versus the climate crisis in the USA, *"is to file lawsuits on the US President for his executive orders in respect to the climate crisis".*

Also, fourteen U.S. states including Louisiana and Wyoming filed lawsuits on March 24, 2021, against President Joe Biden's administration, challenging his pause on new oil and gas leasing on federal lands and waters. If this social and economical consciousness characterized the American industry, what to expect from developing countries? Society (humanity) *"must make **SACRIFICES** as governments, institutions, companies or personal human beings, everyone and everybody, to accept and impose a Global Planet (or CLIMATE) Tax. Bisede a carbon footprint of each product, each fossil fuels company around the panet to pay this Climate Tax under UN jurisdiction.* COP26 which will take place in Glasgow UK from November 1 to 12, 2021, has to tackle this issue as an urgency. That tax must be used only to attenuate the impacts of the climate crisis. A different form of this tax is already applied in Singapore. Some owners of apartment buildings are losing ownership of the building after 100 years of use. Those apartments have been built by squeezing the planet of its resources, and are for sale. The government uses the money with parliament approval. That may be a source of the *"Planet Tax"*

Population growth and refugees, *"during the climate crisis, is a very sensitive issue".*

Rapid population growth, lack of access to food and water, and increased exposure to natural disasters mean *"more than 1 billion people face being displaced by 2050,"* according to a new analysis of global ecological threats. Natural disasters including floods, droughts, cyclones, rising sea levels, and rising temperatures, are climate outcome impacts on the population. The world now has 60% less freshwater available than it did 50 years ago, while demand for *"food is forecast to rise by 50% in the next 30 years".* Those factors, combined with natural disasters mean even stable states are vulnerable by 2050. Climate refugees fall in the same category as population growth. About *"10.3 million people were displaced by climate change-induced events"* such as flooding and droughts *"in the last six months (up to March 2021)"*, the majority of them in Asia. (Last data shows that: Globally, *"17.2 million people were displaced in 2018 and 24.9 million in 2019").* Some people believe that the world is too crowded. Keeping control of population growth could be an answer to tackling climate change. Many experts disagree.

"Is population an issue in climate change? Absolutely. Is it underreported, underrated, and under-talked-about as an issue in climate change? Absolutely. If it were just Adam and Eve on the planet, they could fly a 747 around the world 24/7 and heat modern Castles, and 25 other homes with coal, and it wouldn't make a difference." And it is true, slowing population growth alone won't *"solve climate change".* Concern about overpopulation has been rather long-standing. Population growth would eventually surpass our ability to provide sustenance for the masses. We don't need to eliminate half of the universe's population to end suffering, such as starvation. We need to educate people in undeveloped countries, that women do not have 7 kids, and see them die of starvation. Today's generation hopes that their kids will have a better life. This should be the norm of life (6). The other thing

to take into consideration about our growing population is that the issue isn't so much about births, but rather about dying. Or more accurately *not* dying. Today, people are living longer, especially in developed countries. *"The key to achieving slower population growth is best done through a rights-based approach that includes educating girls and providing universal access to family planning and reproductive health services.*

Education. *The education of the masses has an enormous role in the human attitude toward climate change"*. Humans have to know how to spot the different ways that are being used, by fossil fuel companies, to convince you to delay action on climate change (See education paragraph 6.2.5. Denial). These companies are using 5 types of denial which are: science, economic, humanitarian, political, and crisis denials. Knowing and understanding these techniques will help fight denials and stupidity against global warming. The big oil, gas, and coal companies want to deflect the attention of the public from the real issue of the climate crisis, knowing that *"they should invest in the crisis, at least $3.5 trillion per year for the next 30 years, to attenuate this crisis"*. Also, I will present some common examples from day-to-day experience. We do not need to use a 7 bedroom 9 bath house in which 2 to 4 people live? and heat or cool the entire house? How many people are living in King's size palaces? How much GHG was emitted to build these palaces? Better transform them in museums, and collect some money for Planet Tax. Same for big mansions all over the planet.

I've seen lightbulbs of 1000 watts lighting all day in front of a business building in Bucharest, or busses with engines turned on an hour before leaving the station. And this example goes on. We have to radically change how we are used to thinking and living. For instance, on a Romania TV news channel one parliament member is asking the government: We are the richest country in Europe in natural gas and 40% of the population doesn't have installed natural gas? On the other side of the planet in California, the governor proposes: all new buildings are provided with electrical outlets, no gas installation is permitted. Here we have a mass education issue. Governments must take actions to educate the masses on global warming, and the entire society has to consider the *"professional, political, social, and ethical" consciousness in their life"*. Without education, humanity can not reach the desired level of consciousness. *"We do education based on IQ, and based on education we reach a higher level of consciousness"* (6).

Risks during a crisis. The scientists emphasized some aspects of the risks society is facing *if* a catastrophe is hitting us? If such a predicament occurs, community intelligence would be crucial. No single person can handle the science implemented in a TV set. The TV set includes a combination of several technologies. If people are isolated in a remote area, it will be difficult for them to employ any basic survival tasks somewhere on the planet. Therefore handbooks for survival or survival guidelines should be prepared to be widely dispersed. To assess and minimize global risks, the international community has to do more in the area of planning. The planning has to be globally, and for the long term, *"done by an International Organization (United Nation) with binding power"*. Nations have to give up some degree of sovereignty to a new global organization for regulation in an area like global warming, the planning of water sources, energy generation, AI development, and cosmic space peaceful usage. All these can be achieved by national governments under citizen's pressure. The young generation who are more conscious, and aware of survival to the end of the century and beyond, are more concerned with long-term risks, especially global warming. They have to engage all segments of society and put pressure on the authorities to take decisive action on the climate crisis (6).

Education and Urgency. If I am going on the streets of San Francisco, which I know pretty well living there for more than 25 years, and ask 100 adult people: What do they know about climate change or global warming? You readers will be surprised by the answers I get. Ahh, I heard about it, it's ok, or conspiracy theories man, forget about it. Mass media is part of educating the public about a climate crisis we live in and we don't know. Many articles about this issue, and well documented, are from Bloomberg Co. in New York. I called them to ask permission to use their articles in my book. The answer was simple: Sir not more than 10% of data from each reference. I am talking about an Emergency of life on the planet Earth, as the United Nations has declared plus 11 more countries *"(Swiss included it in their Constitution)"*, and you Bloomberg Co., instead of being honored by my action, you constrain your information on a subject to educate the public, and be top of the online news for decades if not centuries to come. I make this pledge for all journalism around the planet.

Finally, on June 3, 2021, I found an article that makes me and everybody reading about it: proud of real, honest, and targeted journalism.

I intend to make a summary of this article. I Hope CCNow will agree with me.

A handful of major newspapers are paying attention to global warming. But most news coverage, especially on television, continues to underplay the climate story, regarding it as too complicated, or disheartening, or controversial. Last April 2021, we (a group of news journalists CCNow) asked the *"world's press to commit to treating climate change as the emergency"* that scientists say it is; their response was dispiriting.

We created Covering Climate Now (CCNow) in April 2019 to help break the media's climate silence. Since then, Covering Climate Now has grown into a consortium of hundreds of news outlets reaching a combined audience of roughly 2 billion people, and the climate coverage of the media as a whole has noticeably improved. But that coverage is still not going nearly far enough. To convey to audiences that civilization is literally under attack, news outlets should play the climate story much bigger, running more stories, especially about how climate change is increasingly affecting: economics, politics, and other spheres of life, and running those stories at the top, not the bottom, of a homepage or broadcast. News reports should also speak much more plainly, presenting climate change as an imminent, deadly threat.

This message is muted at best today, and the result is predictable. In the United States, only 26 percent of the public is "alarmed" about climate change, according to opinion polls analyzed by the Yale Project on Climate Change Communications (a member of the CCNow consortium). One reason why? Less than a quarter of the public hear about climate change in the media at least once a month. Good journalism leads the conversation, and there is certainly plenty of climate news worth covering these days. In a pair of stunning developments last week, a court in The Netherlands ordered the Royal Dutch Shell oil company wants to reduce customers' greenhouse gas emissions to be 45% by 2030 and not by 20%, in accordance with the Paris Agreement, even as shareholders of ExxonMobil and Chevron rebelled against management's refusal to take strong climate action. A week earlier, the International Energy Agency (IEA) declared that all new fossil fuel development must stop to prevent irreversible climate destruction. The climate emergency is upending politics, economics, and virtually every other subject journalists cover, and newsrooms need to catch up. They can start with the Climate Emergency Statement that CCNow issued in April as part of our Earth Day coverage. Co-signed by eight of our partners: Columbia Journalism Review: *The Nation, The Guardian*, Scientific American, Noticias Telemundo, *La Repubblica, The Asahi Shimbun*, and Al Jazeera English. The statement's first sentence said, "it is time for journalism to recognize that the climate emergency is here." Emphasizing that this was "a statement of science, not politics," the statement linked to articles in peer-reviewed journals where thousands of scientists affirmed that fact. The statement noted that the COVID-19 pandemic illustrated how well news outlets can cover emergencies when they commit to it, and it invited journalists everywhere to apply that same urgency to the climate story. More than 30 newsrooms have now signed the statement, but some major outlets told us privately they won't sign. The phrase "climate emergency" sounded like activism, they said; endorsing it might make them look biased. Instead, they added, they would let their climate coverage speak for itself. But that's the problem. Their coverage does speak for itself, and it is simply not reflecting the facts of the reality. It is a fact that thousands of the world's scientists, including many of the most eminent climate experts, say humanity faces a climate emergency. Most major news outlets still present climate change as no more important than a dozen other public issues, when the fact is that if the world doesn't get it under control, fast, climate change will overwhelm every other issue. Another fact: the *"climate emergency comes with a time limit"*, wait too long to halt temperature rise and it becomes too late. We're not obsessed with whether a news outlet does or doesn't use the term "climate emergency"; what matters is whether the outlet's overall coverage treats climate change like an emergency. For example, does the outlet give the climate story the same 24/7 coverage it has devoted to the COVID-19 pandemic or, before that, the 9/11 terrorist attacks, or other landmark events? Has it reoriented its newsroom and reassigned reporters to cover the climate story? Do its journalists present the story with a sense of urgency?

At an IPCC summit in Glasgow, November 1 to 12, 2021, world leaders are supposed to adopt much stronger measures against the climate emergency. Journalists have a responsibility to make sure the public understands what's at stake with global warming. Crucially, humanity already has the technologies and solutions to decarbonize our economies; what's needed is the political will to implement them. Journalists also have a responsibility to hold powerful interests accountable for doing what's needed to preserve a livable planet. That starts with telling the truth: about the climate emergency, its solutions, and how little time remains before it's too late. I do believe that this short paragraph of "journalism education of the public," is key for other news media, not to restrain their publications on Climate Change to any percentage.

What we do in the next 10 years, may determine the future of all life on the Earth.

"This falls on anyone's shoulder, to change our global warming and to act now." Collapseologists are warning humanity that business-as-usual will make the Earth uninhabitable. But the way governments and big fossil fuel companies are acting, shows that humanity is going to disappear. I still hope that somewhere is a gram of conscience and we can be saved.

Prof. Theodore Vornicu, Ph.D.
Marbella, July 04, 2021, SPAIN

References

1. Yahoo Finance, By: Tom Belger, Jan. 19, 2021
2. The Guardian, By: Bill Mc. Kibbeh, Nov. 3, 2020
3. Yahoo News, By: David Knoles, Nov. 24, 2020
4. Bloomberg, By: Seijel Lishan, Feb. 8, 2021
5. Stateimpact Pennsylvania, Anne Danahy, Jan. 14, 2021
6. Theodore Vornicu, Humanity Must Survive the 21st Century, Amazon.com, USA, UK, Fr. De. Es. It. Jp. Aust. June 2019.
7. Reuter, By: Mathew Green & James Spring, Dec. 15, 2019.
8. YAHOO LIFE, BY: Deepak Chopra, May 4, 2021
9. Reuter, By: Kate Kelland, July 23, 2020
10. Kai-Fu Lee, AI Super Powers China, Silicon Valley, and the New World Order. Houghton Mifflin Harcourt Publishing Company, 2018 New York Boston.
11. Martin Rees, On the Future, Prospect of Humanity, Princeton University Press, Oxford 2018.
12. Time, By: Justin Worland, July 16, 2020.
13. Yahoo News, UN - COP25, 2019 Madrid.
14. Wikipedia
15. Guardian, By: Bill Mc Kibben, Nov. 19, 2019
16. Conversatio, By: Wolfgang Knorr, Apr. 16, 2020
17. Quartz, By: Tim Mc Donnel, July 4, 2020
18. Rolling Stone, By: Jeff Goodell, March 27, 2020.
19. USAToday, By; Doyle Rice, Nov. 12, 2020
20. USAToday, By: Doyle Rice, Apr. 16, 2020.
21. NBC News, Denise Chow, Colorado, Aug. 10, 2020.
22. AFP, By: Giovanna Fleitas, Apr. 5, 2020
23. AP, By: Andrew Meldrum, Nov. 17, 2019.

24. The Independent, By; Louise Hall, June 30, 2020.
25. Camp Fire. Everybody at some point was certain they were going to die.
    By: Louise Boyle, Nov. 27, 2020.
26. Bloomberg, By: Edward Johansen, Jan. 4, 2020.
27. Bloomberg, Jason Scott, De. 18. 2019
28. HuffPost, By: Mary Padenfus, Aug. 16, 2020.
29. Guardian, By: Oliver Milman & Vivian Ho, Oct. 7, 2020.
30. Yahoo Deadline, By: Tom Tapp, Sept. 12, 2020.
31. USAToday, Staff, Aug. 6, 2020.
32. HuffPost, By; Nina Golgowsky, Aug. 19, 2020,
33. The Independent, By: Louisa Boyle, Sept. 17, 2020.
34. Yahoo Stuff, Blaze in Sonoma & Napa, Sept. 15, 2020
35. Deadline, By: Tom Tapp, Sept. 11, 2020
36. BBC, Amazon Fire, Oct. 2020.
37. AFP. By: Louis Ganot, Sept. 11, 2020.
37.a. Yahoo News, Indonesia Fire, Nov. 15, 2020.
38. Bloomberg, Susie Neilson, Sept. 10, 2020.
39. Business Insider, By: A. Woodward, Aug. 21, 2019
40. Yahoo News, By: Jasmine Alguilera, Aug. 3, 2019.
41. Yahoo News, By: Cody Weddle, Nov. 24, 2020.
42. Yahoo News, By: Gina Martinez, Sept. 14, 2019
43. AP. By: Ramon Espinosa, Sept. 2, 2019.
43.a. Yahoo News, Hurricane Derecho, Aug. 10, 2020.
44. BBC Stuff, Aug. 24, 2020.
45. BBC Stuff, Oct. 10, 2020.
46. Yahoo News, Central America Storms, Nov. 2020.
46.a. Yahoo News, Iota Hurricane, Nov. 24, 2020.
47. The Week, By: Catherin Garcia, Sept. 17, 2020.
47.a. BBC, Cyclone Amphan, May 20, 2020
48. The Telegraph, By: Nicola Smith, Nov. 12, 2020.
48.a. Yahoo News, Typhoon Maysak, 2020.
49. Yahoo News, Molave Typhoon, Oct. 28, 2020.
50. Guardian, By: Oliver Milman, APR. 16, 2020.
51. USAToday, By: Doyle Rice, July 20, 2020.
52. Yahoo News, "California's Trillion-Dollar' Mega Disaster

No One is Talking About", By J. Schlosberg, Dec. 4, 2020.

53. The Independent, By: Graig Graziosi, Feb. 17, 2020.

54. Reuter, By: Maria Cospan & Rich McKay, May 20, 2020.

55. USAToday, By: Doyle Rice, July 15, 2020.

56. Reuter, Chinese Dam Collapse, July 20, 2020.

57. BBC Stuff, Severe Flooding Engulf Eastern China, July 13, 2020.

58. AFP Stuff, Hundreds of Rivers Swell in China Summer Floods, July 13, 2020

59. Bloomberg, By: Anjani Trivedi, Aug. 28, 2020.

60. The Telegraph, By: Abbie Cheeseman, Sept. 6, 2020.

60.a. Yahoo News, Polar Vortex, Feb 2021.

61. Oilprice.com, By: Charles Kennedy, June 12, 2020

62. Yahoo News, Libya Oil, 2020.

63. Yahoo News, India oil developments. 2020.

64. IEA Report, Oil stand past Pandemic, Oct. 2020.

65. Oilprice.com, By: Irina Slav, Feb. 2, 2020

65.a. Yahoo News, Texas Flaring, 2020

66. Oilprice.com, By: Mathew Smith, July 22, 2020.

67. CBS News, By: Andrew McNamara, Jan. 2020.

67.a. Cristian Science Monitor, By: Ludmila Lohan, May 2020.

68. Yahoo News, UK Electricity, May 2020.

68.a. AP, By: Kevin Tablet, May 31, 2020.

69. AP, Frank Jordanas, July 3, 2020.

70. Yahoo News, Spain & Portugal to phase out coal. Jan. 22, 2020.

71. Fax Business, By: Brittany de Lea, March 23, 2020.

72. Associated Press, By: Christina Larson, Dec. 2, 2019

73. Wikipedia

74. USDA, Regenerative Agriculture Report, 2020.

75. The Telegraph, By; David Perry, Oct. 15, 2019.

76. Reuter, By: McKinsey Consultancy, May 5, 2020.

77. Oilprice.com, By: Irina Slav, July 27, 2020.

77.a. Yahoo News, Earth Day 50 Years Anniversary, April 22, 2020.

78. Los Angeles Time, Board Editorial, Oct. 19, 2020.

79. Tech Crunch, By: Danny Crichton, Oct. 22, 2020.

80. Los Angeles Time, AIAC, By; Debra G. 2020.

81. Theodore Vornicu, Bauphysik Magazine, Jan. 1988, Germany, Energy Conservation in Buildings.
82. Yahoo News, By; RashelK Beals, Nov. 1, 2019.
83. Yahoo News, By: Lawrie Wright, 2020.
84. Yahoo Finance, By: Eva Krukowska, July 1, 2020.
85. USAToday, By: Brinley Hineman, Oct. 12, 2020.
86. Oilprice.com, By; Editor oilprice.com, July 13, 2020.
87. Reuter, Kile McLellan, Feb. 3, 2020.
88. Los Angeles Time, By: Samuela Masunaga, Oct. 19, 2020.
89. Business Insider, By: Brittany Chang, Oct. 20, 2020.
90. Popular Mechanic, By: Carolina Delbert, Aug. 21, 2020.
91. UN Report on Shipping Decarbonization, Jan. 20, 2020.
92. Yahoo News, By: Jonathan Saul & Nina Chestney, Oct. 30, 2020.
93. Reuter, By: Jonathan Saul, Oct. 28, 2020.
94. Reuter, By: Andy Home, 2020.
95. RNF Report, 2020.
96. AFP. By: AFP Board, Oct. 20, 2020.
97. Yahoo USA, By: Joe Gamp, May 14, 2020.
98. AFP, Board, Wadi's Capital Chokes. Sept. 25, 2020.
99. Yahoo News, Peatland and CO2 in CC Influence., 2020.
100. The Independent, By Louise Boyle, May 11, 2020.
101. The Black Monopoly, Oct. 8, 2020.
102. 1o2. The Independent, By; Herry Cockburn, 2020.
103. The Conservation, Aug. 28, 2020.
104. Bloomberg, By: Eric Raston & Benn Bain, Sept. 9, 2020.
105. Quartz, By: Tim McDonnell, Oct. 21, 2020.
106. Time, Justin Worland, March 4, 2020.
107. Bloomberg, Jess Shankleman, Sept. 1, 2020.
108. Reuter, By: Roslan Khasawneh, Sept. 25, 2020
109. The Week, By; Mathew Walther, Nov. 23, 2020.
110. Yahoo News, European Pollution Cost $190 Billion/year, 2020.
111. Reuter, China says, the environmental style is grim despite 5 years of progress. Oct. 21, 2020.
112. Yahoo News, China Must Commit to EV Battery Recycling. 2020.

113. Step Price, By: Yomi Kazeem, Sept 25, 2020.

114. USAToday, By; Jorje Ortiz, Aug. 20, 2020.

114.a. Yahoo News, Entire West Coast With the Worst Air Quality on Earth, Sept. 11, 2020

115. Quartz, By: Zoe Schlanger, Aug. 2019.

116. Bloomberg, By: R. Adams Heard, Oct. 8, 2020.

117. The Telegraph Editor, By: Yreth Rosen, Oct. 21, 2020.

118. USAToday, By: Doyle Rice, Feb. 13, 2020.

119. Bloomberg, By: Heesu Lee, Feb. 12, 2020.

120. Bloomberg, By: Frances Schwartzkopff, Oct. 30, 2020.

121. Yahoo News, By: Ellen Knickmayer, March 24, 2020.

122. Microsoft Climate Innovation Fund.

123. Oilprice.com, By: Irina Slav, Sept. 14, 2020.

124. Yahoo News, By: Brain Cheung & Valentina Caval, Feb. 19, 2020.

125. Market Watch, By: Rachel K Beals, Oct. 19, 2020.

126. Fortune, By: Daniell Pinto & Ashley Bacon, Oct. 16, 2020.

127. CoinDesk, By: Ian Allison, March 4, 2020.

128. Oilprice.com, By: Leonard Hyman & William Tilles, Jan. 18, 2020.

129. Yahoo Finance, By: S. Kishan, & Sydney Maki, Oct. 14, 2020.

130. Bloomberg, By: Dan Murtaugh, June 17, 2020.

131. Reuter, By: Simon Jesopp, Nov. 18, 2020.

132. Bloomberg, By: Dina Bass, July 21, 2020.

133. Yahoo Financing, By: Reggie Wade, Feb. 17, 2020.

134. Bloomberg, By: Greg Geizinis& Graham Steele, Nov. 19, 2019.

135. Yahoo Financing, By: Oscar Williams-Grut, Jan. 27, 2020.

136. Tutor Business Editor, How California's wildfire can spark a financial crisis, Sept. 2020

137. Reuter, By; William Grut, Sept. 2, 2019,

138. Yahoo News, By; Kate Abnett, Nov. 6, 2020.

139. British Petroleum, By: Bernard Looney CEO BP, Report, June 2019.

140. HuffPost, By: AmanDa Shupak, Aug. 19, 2020.

141. Technology Ideas, By: Peter R. Orszag, Sept. 22, 2020

142. Oilprice.com Editor, By: Dec. 13, 2020.

143. Yahoo Finance Editor, 2020.

144. Smarter Analyst Editor, Sept. 10, 2020>

145. Energy Global, By: Bella Watch, Dec. 16, 2020.

146. Zack Equity Research, Sept. 29, 2020.

147. 147 Oilprice.com Editor, Nov. 23, 2020.

148. Yahoo News, By: Millanjan Banerjee, Aug. 20, 2020.

149. Oilprice.com, By: Irina Slav, Nov. 9, 2020.

150. Oilprice.com, By: Hally Zaremba, July 15, 2020.

151. Reuter, By: Nicola Broom, Dec. 15, 2020.

152. Reuter, By: Sonali's Paul, Dec. 11, 2020.

153. 153a. Oilprice.com, By: Haley Zaremba, Oct. 1, 2020.

154. AP., By: Keith Redler, Nov. 11, 2020.

155. Energy Global, By: Bella Weetech, Dec. 15, 2020.

156. The Week Editor, Dec. 13, 2020.

157. Bloomberg, By: Laura Millan, June 10, 2020.

158. Yahoo News, By: Matthew Dilallo, Sept. 27, 2020.

159. Yahoo News, Natural Gas Hydrogen Injection, 2020.

160. Yahoo News, Hydrogen Delivery to Transportation, 2020.

161. AccessWire Stuff, Dec. 10, 2020.

162. Reuter Stuff, Sept. 22, 2020.

163. In The Known, By: Emerald Pellot, Oct. 1, 2020.

164. Popular Mechanics, By: Carolina Delbert, Sept. 9, 2020.

165. AFP. Editor, Aug. 19, 2020.

166. Oilprice.com Stuff, Sept. 27, 2020.

167. National Review, By: William Levin, Dec. 15, 2020.

168. E&E News, By: Nathan G, July 24, 2019.

169. BBC News, By: Paul Rincon, Oct. 29, 2020.

170. Oilprice.com Editor, Sept. 28, 2020

171. Korea Artificial Sun, By: National Research Center for Science and Technology. 2020.

172. Oilprice.com, By: Alex Kiman, Sept. 16, 2020.

173. The energy on Earth, By: David Roberts, Oct. 21, 2020.

174. Benzinga, By: Jocee Tenn, July 14, 2929.

175. InvestorPlace, By: Luke Longo, Oct. 16, 2020.

176. World Environment, Sept. 13. 2020.

177. BBC By: Helen Brigs, Jan. 27, 2021.

178. USAToday, By; Doyle Rice, Ja. 17, 2020,

179. Nature Communication Journal, 2020.

180. Reuter, By: Mioara Warburton, Aug. 7, 2020.

181. !81. Yahoo News UK, By: Roh Waugh, Oct. 19, 2020.

182. AFP. The study, Oct. 16, 2020.

183. The Conversation, By: Ann Rowan, Aug. 3, 2020.

183.a. Yahoo News, Himalaya Tsunami, Aug. 3 2020.

184. The Conversation, Editor, July 17, 20.

185. NBC News, Editor, Feb. 18, 2020.

186. BBC Yahoo News, Editor, May 14, 2020,

187. The Guardian, N.Y. By; Oliver Milman, Aug. 4, 2020.

188. The Conversation, By; Chole Brimicombe, Aug. 20, 2020.

189. The Guardian, By: Erin McCornick, Sept. 4, 2020.

190. News 18, By: Santosh Chavbev, Sept. 1, 2020.

191. Reuter, By: Megan Rawling, 2020.

192. The Conversation, By: Harriet Ingleaug, AUG. 7, 2020.

193. Yahoo News, June 20, 2020.

194. Reuter, By: Thin Lei Win, Sept. 16, 2019.

195. Reuter, By; Thin Lei Win, Nov. 28, 2019.

196. Time, Area Chen, May 20, 2020.

197. Reuter, By: Thin Lei Win, March 20, 2020.

198. AP Editor, Oct. 6, 2020.

199. Yahoo News UK, Jan. 20, 2020.

200. The Conversation, Editor, April 20, 2020.

201. Reuter, By: Luke Baker, Sept. 9, 2020.

202. CBC News, Editor, Oct. 25, 2019.

203. United Nation Security Council, 2020.

204. The Guardian, By: Jojo Metha & Julia Jackson, Feb. 02, 2021.

204.a. International Criminal Court-UN.

205. Oilprice.com, By: Haley Zaremba, Oct. 22, 2019.

206. Tech - Crunch, By: Jonathan Shieber, Feb. 15, 2021.

207. Yahoo News, By: Alexander C. Kaufman, Jan. 22, 2021.

208. Tech - Crunch, By: Valerie Volcovici, Jan. 19. 2021.

209. AP. By: Mike Corder, Jan. 22, 2021

210. Yahoo News, DAC (Direct Air Capture)) Technology, 2020.

211. Axios, By: Bryan Walsh, Feb. 17. 2021

212. Yahoo News, By: Elana Dure, Jan. 7, 2021.

213. Business News, Editor, Feb. 3, 2021.

214. Tech - Crunch, By: Jonathan Shieber, Feb. 2, 2021.

215. Politico, By; Zack Colman, Feb. 16, 2021.

216. BBC News, By: Justin Rowlatt, Dec. 2, 2020.

217. USAToday, By: Doyle Rice, 2021.

218. Yahoo Finance, By: Akiko Fujita, Jan. 28, 2021.

219. Reuter-Tech Rules, By: Sabine Siebold & Kate Abnet, Feb. 19, 2021.

220. Yahoo News, CCS Carbon Capture & Sequestration), Issues, 2020.

220.a. Business Insider, By: Alin Woodward, Nov. 13, 2020,

221. Los Angeles Time, By: Jacques Leslie, Jan. 4, 2021.

222. The Outside The Box, Feb. 8, 2021.

223. BBC News, By: Sharanjit Leyl, Feb. 5, 2021.

224. Reuter, By; Simion Jessop, Feb. 17, 2021.

225. AFP, By: Patrick Galey, Nov. 10, 2020.

225.a. Yahoo News, Earth Day 2021, Abstract Report.

226. FirstPerson, By; Alexander C. Kaupman, Nov. 18, 2020.

227. People Magazine, By: Jonathan Scott, Nov. 2020.

228. David Page, Practical guide to living with Climate Change, Feb. 8, 2021.

229. BBC News, By: Matt McGrath, Dec. 9, 2020.

About the Author

Prof. Theodore Vornicu, San Francisco, CA. By Mircea Boita

Prof. Theodore Vornicu, Ph.D., has written textbooks for graduate students and more than thirty articles focusing on engineering science. He has specialized in structural building research and design for earthquake-prone areas and has taught at universities in Romania and California. He earned his Ph.D. diploma from Gheorghe Asachi Technical University of Iasi, Romania, and an honorary degree as associate professor from the same university.

Lightning Source UK Ltd.
Milton Keynes UK
UKHW012334250821
389480UK00001B/47

9 781665 592420